Managing Organizational Knowledge

3rd Generation
KNOWLEDGE MANAGEMENT
...and Beyond!

Managing Organizational Knowledge

3rd Generation
KNOWLEDGE MANAGEMENT
...and Beyond!

Charles A. Tryon, Jr.

CRC Press is an imprint of the
Taylor & Francis Group, an **informa** business

CRC Press
Taylor & Francis Group
6000 Broken Sound Parkway NW, Suite 300
Boca Raton, FL 33487-2742

© 2012 by Charles A. Tryon, Jr.
CRC Press is an imprint of Taylor & Francis Group, an Informa business

No claim to original U.S. Government works

Printed in the United States of America on acid-free paper
Version Date: 20120125

International Standard Book Number: 978-1-4398-8235-1 (Paperback)

This book contains information obtained from authentic and highly regarded sources. Reasonable efforts have been made to publish reliable data and information, but the author and publisher cannot assume responsibility for the validity of all materials or the consequences of their use. The authors and publishers have attempted to trace the copyright holders of all material reproduced in this publication and apologize to copyright holders if permission to publish in this form has not been obtained. If any copyright material has not been acknowledged please write and let us know so we may rectify in any future reprint.

Except as permitted under U.S. Copyright Law, no part of this book may be reprinted, reproduced, transmitted, or utilized in any form by any electronic, mechanical, or other means, now known or hereafter invented, including photocopying, microfilming, and recording, or in any information storage or retrieval system, without written permission from the publishers.

For permission to photocopy or use material electronically from this work, please access www.copyright. com (http://www.copyright.com/) or contact the Copyright Clearance Center, Inc. (CCC), 222 Rosewood Drive, Danvers, MA 01923, 978-750-8400. CCC is a not-for-profit organization that provides licenses and registration for a variety of users. For organizations that have been granted a photocopy license by the CCC, a separate system of payment has been arranged.

Trademark Notice: Product or corporate names may be trademarks or registered trademarks, and are used only for identification and explanation without intent to infringe.

Library of Congress Cataloging-in-Publication Data

Tryon, Charles A.
  Managing organizational knowledge : 3rd generation knowledge management and beyond / Charles A. Tryon, Jr.
       p. cm.
  Includes bibliographical references and index.
  ISBN 978-1-4398-8235-1 (pbk. : alk. paper)
  1. Knowledge management. 2. Information technology--Management. 3. Information resources management. I. Title.

HD30.2.T82 2012
658.4'038--dc23   2011049809

Visit the Taylor & Francis Web site at
http://www.taylorandfrancis.com

and the CRC Press Web site at
http://www.crcpress.com

To my wife, Tresa.
The love of my life.
To my daughters, Amanda and Casey.
You bring me so much joy.
To the memory of Ashley Rhea.
Never forgotten.
To my mother and father.
The leaders of our band.

Contents

Acknowledgments xi
Introduction xv

Chapter 1   Knowledge as an asset—Really? 1

Chapter 2   The new realities of knowledge management 5
The growing knowledge gap 6
   "Boomer" retirements 6
   Downsizing 7
   Unforced resignations 7
   Internal promotions 7
   Market globalization 8
   Technology advances 8
   Business complexity 8
Knowledge opportunities 8
Return on investment 9
A call to action 11

Chapter 3   KM beliefs 13
Knowledge sharing and reuse 15
Learning organization 16
Best practices 17
Communities of practice 19
Conclusion 20

Chapter 4   KM processes 21
Knowledge discovery 21
Knowledge capture 22
Knowledge organization 23
Knowledge use 24
Knowledge transfer 26
Knowledge retention 26
Conclusion 28

Chapter 5 Defining organizational knowledge 31
Knowledge categories 32
   Individual knowledge 32
   Global knowledge 32
   Organizational knowledge 33
Explicit, tacit, and implicit knowledge 34
   Explicit knowledge 34
   Tacit knowledge 34
   Implicit knowledge 35
Knowledge characteristics 37
Why bother? 38
Conclusion 39

Chapter 6 Recognizing organizational knowledge 41
Data 42
Information 43
Decision making 44
Conclusion 45

Chapter 7 The knowledge retention policy—Level one 47
General management statement 48
Knowledge asset inventory 49
   Knowledge areas 50
   Knowledge topics 51
   Knowledge expert 53
   Organizational importance 53
   Transfer status 54
   Knowledge transfer mechanisms 54
KRP activities 56
Conclusion 59

Chapter 8 The knowledge retention policy—Level two 61
Documentation 62
Training 63
Apprenticeships 64
Mentoring/coaching 65
Cross-training 65
Communications 65
Conclusion 66

Chapter 9 A model for managing organizational knowledge 67
KIPPAR Model 67
The knowledge inventory 68
The artifacts pillar 69

The processes pillar 70
The projects pillar 71
Repository products.. 72
KM or ECM 73
Conclusion... 74

Chapter 10 Implementation strategies..**75**
KM initiation activities.. 75
    Establish common definitions 75
    Define your knowledge management vision 76
    Assess organizational beliefs 77
    Encourage communities of practice 78
    Launch your KM effort as a project... 79
    Build a knowledge portal.. 80
    Create a sample knowledge retention policy 81
KM operational activities.. 81
    Define personal knowledge goals... 81
    Harvest knowledge assets from projects.. 83
    Engage contributing disciplines 84
    Emphasize "management" elements of KM 84
Conclusion... 87

Chapter 11 Knowledge management solutions.......................................**89**
Functionality.. 90
Usability 91
Personalized knowledge apps... 92
Organizational portals... 93
Project portals.. 94
    Common entry point 95
    Distinct presence 95
    Administrative versus functional content.. 96
    Transition artifacts 97

Conclusion **99**
Appendix A: KM Vision Statement... 101
Appendix B: KRP—General Management Statement 103
Appendix C: KRP—Intellectual Assets Inventory.................................. 109
Appendix D: KRP—Knowledge Transfer Details 113
Index 121
About the author 125

Acknowledgments

> "If you would not be forgotten, as soon as you are dead and rotten, either write things worth reading, or do things worth the writing."
>
> **Benjamin Franklin**

Authors often open their books by thanking the many people who contribute to their work. So please pardon this personal indulgence. These names may not be recognizable to you but they have each touched me in a notable way.

Thousands of very bright people have attended my workshops over the years. Although it is impossible to remember or credit each of them, they have stimulated many of the ideas you will find in these pages. They have also helped me understand how a wide variety of businesses work. I hope you guys buy this book and then send me your comments. You honored me simply by being part of my growth process.

I've examined the implications of the knowledge age since the early 1980s, however, my formal research into knowledge management began when I met and collaborated with Dr. Suliman Hawamdeh while he was a professor at the University of Oklahoma. Suliman, thank you for the many hours of stimulating and insightful discussion. You gave me an early forum to explore many of the topics included in this book.

In Chapters 7 and 8, I introduce a knowledge inventory called a Knowledge Retention Policy, or KRP. I created the first KRP for the City of Tulsa Police Department. After being rejected by two clients (too conceptual), Chief of Police Dave Been (ret.) agreed to put his management team at my disposal to see what would happen. I'm sure he didn't really understand what he was agreeing to do but he trusted me. Thank you, Dave. Keep hitting 'em straight.

There were several dozen people engaged in the Tulsa PD effort, but Major Paul Williams deserves the bulk of the credit. He had the insights I needed to identify their knowledge areas and knowledge topics. He also saw the long-term value of the document, even when a new command staff for Tulsa PD did not. By the way, next time you travel through Springfield, Missouri, give Chief Paul Williams a shout. And, yes, there is a correlation to his work on the KRP.

Fortune brought Phillip Barnett and Lenard Jasczak, of PricewaterhouseCoopers, my way. Based on their work creating a KM organization for PwC, they immediately connected with my efforts and have provided encouragement and validation. Thank you for your shared insights.

I am no self-made man. I am especially grateful to a handful of people who shaped my professional skills. They include Ed Yourdon, Tim Lister, and Tom DeMarco. If you recognize that trio of names, you are also a methodology junkie, and as old as I am. To that group of distinguished thinkers I add a great personal friend and mentor, Franklin L. Kastl, III. Frank, you have benefited the lives of so many people. You were one of the first people I met in business who understood our responsibility to share our knowledge. You led by example.

To my family, thank you for your patience and helping me to become the person I am. My parents, Drs. Charles and Lottie Tryon, Sr., inspired me to tirelessly pursue education and creative ideas. Mom and Dad, you helped set my life compass by sharing your faith with me. You also displayed an unfailing belief that I would succeed. You deserve much credit for anything I accomplish.

To my wife, Tresa, you have been my best friend for more years than we can imagine. You remain the only love of my life. You and I shared the joy of raising two beautiful daughters, Amanda and Casey. You were also there during the dark days when we lost Ashley. Both the love and the sorrow made me a better person.

My favorite name is now "Papa." Mandy, those three grandkids have redefined my world. No struggle or challenge has been too great that it was not overcome by the hugs, kisses, and wrestling matches with Madeline, Landon, and Cailyn. They are the absolute delight of my days. Nothing can ever replace the memories of holding their young hands or feeling them reach out to me while they sleep. I cannot thank you and John enough. Little ones, you will grow into your own dreams and plans. Although you may not remember much from your youngest years, frozen in my mind are your images at this writing. You will never know how much you encouraged me during some very trying times. Your unconditional love is undeserved and overwhelming. And Mandy, I will forever treasure the special times we shared on your way to adulthood.

Casey, your sense of humor and sarcastic wit has launched many a conversation with, "Have you heard what Casey did?" You are one intelligent, talented, and funny young lady. You make me smile.

Acknowledgments

Thanks also to Dr. David Kendrick who offered me the opportunity to lead an ultimate knowledge project at MyHealth Access Network.

Thank you to those who reviewed this manuscript and encouraged me to keep going, with special recognition and appreciation to Joe Colannino.

Lastly, thank you, reader. For without you electing to ponder this content, it would be a rather futile effort in self-indulgence. Writing these words helps clarify my thoughts, but knowing that others also find value in what I think is a humbling and gratifying notion. Enjoy.

Introduction

"Wake up, Neo!"

(Opening line on computer screen) *The Matrix*

The Matrix is a movie loaded with symbolism and innuendo. In the scene where Neo (anagram for "one" as in "The One") is rescued from the womb (the true meaning for "matrix"), he encounters Morpheus (god of dreams) and inquires why his eyes hurt. "You've never used them before," is the reply. The contents of this book represent my knowledge management (KM) "wake up" call. It describes my venture into a world of the familiar and the new. It gave me a new way to view the value of organizational knowledge. And in some cases, I was using my eyes for the first time.

What I discovered during my journey may be the most significant management challenge facing twenty-first century organizations ... how to capture, transfer, and share meaningful knowledge that is vital to their survival. I found insightful debate along with some serious misdirection. I learned that organizations are investing unimaginable amounts of time and money loading volumes of documents into electronic repositories in the vain hope they will somehow be able to distill critical knowledge from these records.

Based on over 30 years of hands-on work as a consultant and educator, I've been afforded the opportunity to learn how a wide variety of organizations work. I know their challenges and needs. Based on these experiences, I've developed a few new insights for the knowledge management space. So what makes this book different? What makes it worth reading? I believe it will provide you, as it has already done for many others, a new perspective on knowledge management. And it will help you see where KM can take you in the future. Let me explain.

I believe that KM, even given its short life span, has evolved through three distinct eras, and is already entering a fourth. These are not mutually exclusive events, but building blocks. The intent of explaining what I see is to help you identify what your organization has accomplished and then intelligently determine where you plan to go. Without this perspective, decisions on KM strategies could be misdirected and disappointing.

First-generation KM—The technology era

The first generation of knowledge management began to reach prominence in the late 1990s through the joining of a corporate need and emerging technology. The business driver was senior corporate executives seeing peers held legally accountable for shoddy or deceitful recordkeeping. There's nothing like watching someone in a similar situation being taken off to jail to spur a little action. Financial and legal realities demanded enhanced means to capture and store important documents and records. And where there is a demand, there will be vendors. Armed with repository technology that could now capture far more than the structured databases of the 1970s and 1980s, solution providers arrived with talk of taxonomies, capture technology, metadata, and search engines. It was called document and records management or DRM. It was also the birth of knowledge management.

This early version of KM remains the way many people still view the discipline. And if all you are looking for is a way to structure and store electronic versions of your documents and records, DRM may still be your answer. But that is not what this book is about. Oh, I talk about technology-based repositories along with the concepts behind taxonomies and metadata, but KM must become far more than document management, or as many refer to it today, enterprise content management or ECM.

Although there is proven value in DRM/ECM, the challenge with this first generation of KM solutions is actually found in the very term "enterprise." Attempting to do anything that crosses an entire organization brings with it a plethora of problems. First, is the complexity of implementing organizationwide initiatives. Mustering the organizational energy, cooperation, and resources needed to tackle these types of initiatives is simply overwhelming for even the most dedicated. Second, most of the benefit from the initial generation of KM is seen only by a handful of people who are charged with protecting the careers of senior executives. Little value trickles down to the operating levels of the organization. As with most distasteful but necessary acts, DRM/ECM is tolerated, but not admired. Perhaps the greatest problem with first-generation KM thought is that senior executives, once again, viewed this as a simple "buy it" solution. They continue to live in a fantasy world where if they throw sufficient funds at a problem, it will be solved. This approach may have worked during the acquisition efforts of the industrial age, but it is not serving the needs of the knowledge age.

Another downside to this "store it and ignore it" era of KM is the resulting organizational structures required to support the global effort.

Due to the granularity and breadth of the massive captured content in these enterprise solutions, it is common to find "knowledge organizations" comprised of dozens or even hundreds of people. It doesn't take long before others in the company, wearied with belt-tightening and staff freezes, began to murmur against an organizational group that seems to provide limited real value.

So how much has your organization enjoyed staying SOX (Sarbanes–Oxley) compliant or meeting HIPAA (Health Insurance Portability and Accountability Act) demands? These are the types of industry and corporate mandates that fueled the need for document and records management. No matter how necessary, it is like enduring a root canal or visiting a proctologist. We know the importance, but it doesn't wake us up in the morning with a grin. First-generation KM solutions were a start, but most of us demand a lot more out of our KM experience.

Second-generation KM—The service management era

The technology dependency of the first generation of KM demanded the direct participation of information technology (IT) departments. With that participation, however, came an opportunity. IT organizations have long been criticized for remaining remote from the real needs of their business clients. In first-generation KM, IT groups quickly recognized that the natural capability of the technology-based repository products allowed them to bring a new level of service to internal and external client organizations.

How often have you called a help desk trying to fix a problem with a product or service, only to be chased around a phone tree, looking for anyone with a clue as to how your problem could be resolved? Making matters worse, to save money, many organizations distanced themselves further from their customers by subcontracting this function to companies who know even less about the problem. Ugh! And they wonder why customer satisfaction surveys paint such a dismal picture.

In rides service management. This strategy is built on recognizing that most customer problems are not unique. And if an organization could capture how someone actually fixed the problem for one customer, others could then access that information and apply it to the next troubled client. Capturing the problem conditions and resolutions, along with suggested workflows and scripts, is a valuable offshoot of knowledge management. But again, it is a very limited perspective.

The service management era of KM, by definition, focuses on problem resolution, and typically on only a microcosm of a problem. It results in fragments of solutions and has a difficult time seeing the big picture that could provide significant innovations and improvements. In many ways, second-generation KM aligns with the incremental improvement strategies of statistical process control and industrial engineering. Again, if that is your goal, great! But this is not the book for you. There are seemingly endless certifications and workshops, all given in the name of knowledge management, to help you become more aligned with solving narrow customer service problems. To be sure, these are valuable activities, but we are left needing more. Enter generation three.

Third-generation KM—The deep knowledge era

The purpose of this book is not to invalidate the initial two eras of knowledge management, but to build on them, and to provide a link to the fourth generation, something organizations are already trying to achieve.

Third-generation knowledge management is about getting to the *deep knowledge* of your organization. It is about the intellectual content that walks out the door every time a valuable employee resigns or retires. It is about understanding how major categories of knowledge flow together to represent the total understanding of a product or service offered by your organization. As we show in these pages, this is about delving into thoughts (some documentable and others not) about how your organization actually works, and then making that knowledge available to anyone with a need for it. It is the knowledge required to create innovative new products and services today, and then be the springboard for major revisions in the future.

This goes way beyond the technology emphasis of the first generation of knowledge management and the problem resolution focus of the second. Deep knowledge is where the real work, and promise, of KM lies. This is where we can discover, capture, organize, use, transfer, and then retain valuable organizational knowledge. This takes a new commitment and new processes. The approaches proposed in these pages will help your organization stem the erosion of current knowledge, create new knowledge, and then share it endlessly. This book will help you position your knowledge as a true asset that may be used to expand and grow your enterprise. It is a strategy I call *managing organizational knowledge*.

The news is already full of organizations, including state and federal agencies, shedding experienced senior workers, often the only ones who actually know how things work. When faced with the loss of critical mandated services, the organizations secure massive emergency appropriations in the vain attempt to recover lost knowledge. There always seem to be sufficient funds to bring the wandering knowledge back into the fold but forget trying to find the financial and time resources it takes to prevent the loss in the first place.

That is what this book is about. If your organization is committed to a path that mindlessly allows brain drain, you will not find a magic wand in these pages. This book is for organizations with sufficient vision to get in front of this trend, while there is actually time to do something about it. I show you how to bring your organization into alignment with expectations for knowledge management, how to recognize even fragmented activities that will prove valuable, and then create a vision for what to do next. I introduce processes and templates for creating a knowledge inventory so you will be able to get your arms around collections of this deep knowledge.

To make all this work, I propose a narrower project-centric approach instead of coming at this enterprise-wide. Naturally occurring projects are the greatest users and creators of integrated, detailed organizational knowledge. So instead of launching a specified KM initiative, projects become the perfect environment to apply the KM concepts proposed in these pages. This strategy, although more incremental than enterprise efforts, will yield certified valuable results faster, and do it while facilitating the creation of new products and services. And most significantly, this strategy will position you perfectly for what I believe will be the fourth generation of KM.

Fourth-generation KM—The personalization era

For many organizations, the future is now! As any knowledge manager can tell you, actually getting people to find and use accumulated intellectual assets remains one of their greatest challenges. There are success stories where companies invested heavily in KM initiatives over an extended period of time and have created massive, useful knowledge portals that their employees apply daily. I've seen these results and I am impressed. But few organizations have the vision or fortitude to travel this path.

On the surface, the more incremental, project-centric approach I propose would seem to contradict building a consolidated integrated view of an organization's knowledge. That is not the case. You can, and should, create a version of the proposed knowledge inventory showing how your organization's knowledge, not people, should be structured. And if you have the resources, you should create a global knowledge portal to support it. But I think a different reality is already upon us.

Imagine one of your salespeople preparing for a series of client visits. Or consider an engineering team heading off to a remote work location. Instead of trying to take the total corporate knowledge repository with them, they simply select and download the knowledge they will most likely need. This may be comprised of a recommended, proven set of tools that are supplemented by personal favorites. Should they run into a situation where they require more, they log in to obtain additional knowledge components from a common pool. When they find a specific expertise is needed, they use a people finder that helps them locate the most likely internal or external expert who can solve their problem.

If you have a smartphone or tablet computer, you already know how this works. It is the world of "apps." Short for "applications," these incrementally created and nonintegrated windows of knowledge access may be downloaded and then organized based on how individual users plan to apply them. Downloaded apps are placed on menu pages and moved around to fit individual preferences. Apps with limited specific purposes are placed in special areas for the term of the need and then removed. You can always return to the common source for apps and reselect a valuable tool or pick one of the emerging products that are added daily.

Social media tools are providing us with new and better ways to connect by using a network of friends. A controlled use of this human interaction capability within a business environment is providing enterprise-wide access to true knowledge experts, both those internal and external to our organization.

My greatest fear for this fourth era is that organizations are already rushing to apply these emerging technologies without first applying the processes of the third generation. As a result, they focus on fragmented solutions to the wrong problems. To truly solve the needs of this generation, we should first capture the deep knowledge that defines an integrated view of organizational knowledge.

So now you know why I wrote this book. This book is my effort to share my observations and strategies. I am writing this to help you explore the significance of the third and fourth generation of KM. And if this is a new topic for you, it is an excellent place to start. In the early chapters, I answer the most basic questions about KM, questions that have surprising answers, questions I needed to resolve before I could plunge more deeply into this topic.

As you read through the pages of this book, keep in mind that there are many perspectives on knowledge management and there are many voices attempting to bring clarity to this emerging discipline. There are likely several topics in this book that I consider innovative and fresh whereas others may just think them ridiculous and impetuous. I'll attempt to justify my point of view so you can decide.

One reality is undeniable. If you are not already full throttle into knowledge management, time is running out. As you will read, there are factors that will overwhelm even the most impressive organization. It has never been easier, or faster, for established organizations to fail. Change will happen in your organization. The question is whether it will be driven by visionary leadership or the knee-jerk reactions that accompany the fear of organizational demise. There is nothing more powerful than the adrenaline rush from the fear of going out of business.

I have no desire to frighten anyone. This is a book about how to solve a problem we all face. It contains real answers to real problems. I hope it

Introduction   xxi

equips you to contribute workable answers for your organization and the individuals in it.

Want to cut to the chase?

If you are prone to reading the last chapters of a book first and just want to know the end results, here is a handy guide. In this book, you can learn how to do the following:

- Explain the underlying reasons why a fresh approach to knowledge management is needed along with an introduction of basic KM concepts (Chapters 1 and 2).
- Create a knowledge management vision statement for your organization that is based on the KM beliefs and KM processes (Chapters 3 and 4). This statement frames any KM implementation.
- Conduct a KM assessment, also based on the KM beliefs and KM processes, of your organization's readiness to meet the stated KM vision.
- Discuss the foundations of knowledge and distinguish the realities of organizational knowledge (Chapters 5 and 6).
- Define a knowledge inventory (Chapters 7 and 8) for a specific knowledge domain that documents the meaningful explicit and tacit knowledge needed to support that area of your organization.
- Provide important elements of information needed to create a knowledge portal that will make knowledge assets easily accessible to the people of your organization.
- Explain the KIPPAR model (Chapter 9) that will help you integrate a wide range of established elements to support a KM implementation.
- Define my roadmap for implementing KM in your organization (Chapter 10). Of course, you may find it best to understand the pieces of the puzzle before you try to put them together.
- In the final chapter (Chapter 11) I suggest strategies for creating technology solutions to implement what you documented in your knowledge.

So enjoy! And I hope this book offers you a few wake-up calls. Then it is your job to alert the rest of the people in your domain of influence and help them use their eyes, maybe for the first time.

chapter one

Knowledge as an asset—Really?

> The man who does not read good books has no advantage over the man who can't read them.
>
> **Mark Twain**

"Knowledge is our most valuable asset!" You hear that statement from many business leaders. But is it really? If organizational knowledge were such a valuable asset, wouldn't it be handled with greater care?

Physical assets are treated with great respect. Think about the last time your organization chose to purchase a new vehicle, add new equipment, or make a major computer hardware upgrade. You likely spent days or weeks examining your options before making the "buy" decision. You certified that it was a true need, not just a short-term want. Then you invested carefully, making sure you had sufficient budget to pay for the new asset. If you saw some level of risk associated with your acquisition, you likely purchased insurance or bought a maintenance agreement to protect it. To keep these physical assets in working order, you scheduled regular maintenance and then made sure the work was done properly. And when the day came that the asset was no longer needed, you had a plan for disposal.

To keep track of all this activity, you use asset management processes and technology to coordinate the life of the asset. Why? Because the asset is valuable! So, how about your organizational knowledge? Do you have a *knowledge* asset management system? Do you even know what organizational knowledge your organization owns? Do you know where it is or who is the current custodian for it? Have you made sure that knowledge is preserved and that other people know how to find and use it? Do you know where you are at serious risk of knowledge loss? Do you have strategies in place to accurately transfer this knowledge from one generation or location of workers to another?

If you can answer "yes" to these questions, you are in rare company. Many leaders proclaim the importance of their organizational knowledge but that seems to be the end of the discussion. They believe that somehow, knowledge assets are doing just fine on their own. They assume good employees will make sure it is cared for and transferred. And many are discovering just how wrong they are.

It is this reality that finally brings knowledge management to the forefront of many senior executives' minds. First discussed in the late 1990s, KM has matured into a formal discipline that can bring significant value to your organization. So, what exactly is knowledge management? Even surface level research will yield a wide range of definitions. In my work, I define *knowledge management* as applying formally defined, repeatable processes that facilitate knowledge discovery, capture, organization, use, transfer, and retention to a specific domain of intellectual assets (more in Chapter 3).

As you can see from the title of this book, I am particularly interested in organizational knowledge, the knowledge that makes enterprises such as yours work. I define *managing organizational knowledge* as a formal integrated application of knowledge beliefs and knowledge processes (KMBPs) to an organization's data, information, and decision making within a specific subset of an enterprise (much more on this in Chapters 3 through 6). KM is a blending of proven organizational strategies with fresh observations, new processes, and enabling technologies.

One feature that makes this book distinctive is the emphasis on a "project-centric" strategy to feed your knowledge base. Other approaches rely on problems or incidents to trigger the capture of new knowledge. By rigorously tracking and evaluating repeated problems in organizational procedures, practitioners argue that modifications may be made that prevent future errors. Further, this second-generation KM approach provides up-to-date scripts that others may follow when they encounter similar situations. Although this emphasis on correcting recognized problems supports the needs of an end customer or of a product, it also has some limitations.

An incident-centric strategy responds to specific, often isolated problems. When addressing these issues, the true root cause may remain undiscovered. Also, the emphasis on "problems" may cause an organization to miss out on extracting knowledge from major opportunities. That is what attracts me to a project-centric approach.

A project is a planned set of work applied to a specific scope that is intended to produce a desired product or service. Projects may be targeted toward solving a narrow vexing problem, however, they often look at a larger, more complete picture . . . and they create more comprehensive solutions. Projects are typically done once within a reasonable period of time, becoming what can be classified as *single-time efforts*. Projects exist naturally in all healthy organizations and you will recognize them in your organization as initiatives, programs, and special assignments.

Since the mid-1980s, I have taught project management and related disciplines to hundreds of organizations and thousands of professionals. What drew me to the KM discussion was recognizing that projects are the most significant source of new knowledge in every organization.

This new knowledge is not contrived but is typically the natural by-product of project activities. Innovative projects are also the most likely users of established knowledge. We use existing knowledge on a project to refine current products and services as well as to create something totally new. Leveraging this knowledge to support a KM initiative simply makes sense.

Unfortunately, once a project is over, the newly created knowledge assets are often unrecognized and left to decay in filing cabinets or on shared drives, and never referenced again. How often have you launched a new project that is very similar to something your firm did just four or five years ago, only to find that most of the important knowledge from the earlier effort is no longer available? Equally vexing is having several people, departments, or divisions of your organization recreating established organizational knowledge instead of sharing what they know. Consider the consequences if one of your competitors finds a way to share significant knowledge, eliminating wasted effort and time when bringing new products to market. How will this affect your ability to compete with them and win?

Although the actual role projects play in KM is addressed in the KIPPAR model (described in Chapter 9), the obvious conclusion is to create a culture that mines *knowledge artifacts* from all projects conducted within an organization. Knowledge artifacts are any representation of knowledge that may be stored electronically. They may be captured using technologies that include text, graphics, sound, and video. Each artifact should be evaluated to determine its true worth to an organization. If the knowledge artifact is considered vital or significant to future business needs, now is the time to develop a knowledge retention strategy.

This approach is far more likely to succeed over time than trying to do an enterprise-wide, dedicated KM project. A project-centric approach to knowledge management will allow your organization to harvest knowledge from a natural source with minimal additional effort. And it can be done, if necessary, on individual projects, without larger corporate awareness or blessing.

I don't want to discourage you from launching an organization-wide KM program. In fact, I suggest just that when I explain the role of the knowledge retention policy. This project emphasis is also not designed to devalue studying and resolving the common problems in your organization. But the emphasis on project by-products will provide your organization with an ongoing source of significant knowledge artifacts as long as you are in business. I provide you with specific project-related strategies later in this book.

Before you spend any additional time on this book, I'd like to share with you some core values. Everything in this book is based on them. If you, or your organization, cannot buy into these statements, you won't realize much value from the balance of this material.

- *The value of organizational knowledge can be measured monetarily.* Defining and capturing organizational knowledge requires time and effort, a cost to your organization. Recreating old knowledge represents a cost that could have been avoided if the original knowledge had been captured and updated on a regular basis. Organizational knowledge is a key ingredient for producing new revenue as it enables the creation of new products and services. Without it, organizations quickly lose established capabilities, customers, and cash flow.
- *Employees are the custodians of this valuable asset.* Your employees are the only vehicle for creating and using organizational knowledge. They are the means for transforming this knowledge into marketable results. And when employees leave, they take their knowledge with them. When a senior employee, who truly understands the inner workings of your organization leaves, the cost of replacing that knowledge is often grossly underestimated.
- *Knowledge experts within your organization must be recognized and rewarded for sharing their knowledge with others in the organization.* Failure to recognize this reality results in knowledge workers hoarding knowledge in an effort to gain job security. Knowledge sharing can only be encouraged, not coerced. In an active economy, knowledge workers may take their talents and skills to a competitor where they believe they are more valued. If your people are truly your most valuable resource, it is vital that this message is reinforced with actions, not just words.
- *Senior management must actively support and encourage knowledge sharing and reuse.* A culture of knowledge sharing must be established by senior leadership along with an emphasis on reusing valuable knowledge. Reusing existing knowledge demands an on-going effort to maintain it and make it findable.

This book is based on a blend of academic research, practical application, and real-life examples. Join me in the quest to manage organizational knowledge as a true respected asset. I hope this book gives you a distinct advantage in this endeavor.

chapter two

The new realities of knowledge management

> Everything discussed here has already happened, it is only the full impacts that are still to come. But few, I suspect, have asked themselves, "[W]hat do these futures mean for my own work and my organization."
>
> **Peter F. Drucker**

The knowledge age has fully arrived! Long predicted by numerous authors, including Peter Drucker, Alvin and Heidi Toffler, and Tom Peters, this shift requires fundamental changes in how we think and operate. Market survivability demands that organizations come to terms with a set of new realities.

Part of this shift is an awakening to the significance of organizational knowledge and the role of the knowledge worker. No longer are your employees "cogs" that fit into an industrial engine; they are now the heart and mind of business transformations and operations. They make the daily decisions that determine the success or failure of your business, including if your customers will return.

These factors are forcing modern organizations to find meaningful responses to a series of growing challenges. At the top of the list is the threat of knowledge loss, a loss that has many interrelated sources. When you lose vital knowledge, you inevitably lose established capabilities. It isn't long before lost capabilities result in lost customers.

Further complicating the matter is that many of the early "solutions" intended to prevent the knowledge drain served only to capture documents and records. For most organizations, purchasing electronic document management systems created a morass of unsearchable content. Professor and friend, Dr. Suliman Hawamdeh, is fond of saying that organizations have turned their wikis, repositories, and intranets into dumping grounds for untrustworthy, unreliable, and unsearchable information. To truly solve the challenge of retaining organizational knowledge, you must start by understanding the risks and opportunities that drive the knowledge management discussion.

The growing knowledge gap

There is a growing gap between what an organization needs to know in order to thrive and what their members actually know. This knowledge gap has many causes including "boomer" retirements, unforced resignations, internal promotions, globalization of markets, technology advances, and general business complexity. The key is determining which of these factors will have the greatest impact on you and how your firm can effectively close the gap.

"Boomer" retirements

The baby boomer generation, those born shortly after the end of World War II, represent the largest influx of workers in the history of business. Complicating this factor is that this legion of employees has stockpiled retirement accounts to fund a comfortable postcareer life. This segment of our working population is about to call it quits, en masse, to enjoy a life of reduced stress and labor. If these employees were simply cogs in the industrial machine, it would be challenging enough to replace so many within a short period of time. But the contributions of these knowledge workers are not based on the labor of their hands and backs, but must be measured by the decisions and actions that result from their educated know-how.

Some organizations are finding a double impact from this reality. Sudden growth periods in a company's history, often accompanying a new product or service, trigger mass hiring. Human resource departments know to start the turnover clock when such events occur. Twenty to thirty years later, they can anticipate a mass exodus when the same people, who fueled the expansion, are now ready to leave. Regardless of the cause, replacing knowledge workers is a difficult, costly, and time-consuming process. You don't capture this set of knowledge by simply having these people "write down everything they know" before you host their retirement parties.

When working with a company to understand the significance of this implication, I often ask an assembled group of managers and senior staff how many of them expect to retire within the next five years. Typically hands are raised by half the people in the room. With their hands up, I then instruct the remainder to look around. You see, the problem of the cumulative knowledge loss will belong to those left behind. If you are planning to be with your organization for a while, it will be your challenge to maintain and grow the operation, without the help of the "boomers."

Some of the anticipated retirements may benefit your organization. Pending retirees may be holding up progress by remaining in a rut of their own making. They may be resisting needed change and their seniority makes it difficult or impossible to work around them. However, before you celebrate their pending departure, you need to make sure you fully understand "what" it is that they know and separate it from the clunky "how" they do their job.

Downsizing

In addition to the implications of baby boomer retirements, challenging economic times are forcing organizations to make difficult staff reductions. Faced with pages of red ink, firms often find it easier to cut senior staff members, with salaries two or three times that of a new employee. What organizations fail to calculate are the implications of the missing knowledge the cheaper replacement employees simply don't possess.

The full implications of releasing senior staff are usually not seen for months or even years. Then comes the challenge of re-creating this lost knowledge or finding a way to entice former employees to return on a temporary contract basis. So much for cost savings!

Unforced resignations

As the full impact of knowledge loss becomes apparent within organizations, many employers resort to finding new sources for the knowledge, namely other companies. Employees in your organization with valuable knowledge will become prime targets for placement firms. All it takes is one of your key people to leave and take the corporate directory and knowledge inventory with him or her. When the fear of being unemployed is absent, ambitious people tend to follow the money.

Internal promotions

Equally troubling is an internal trend sure to sabotage well-intended succession planning. Many HR organizations encourage the promotion of known internal candidates to fill positions left open for one of the reasons listed above. For younger employees, this may be one of the greatest opportunities they will ever see for accelerated internal advancement. But this plum will bring its own price. As these newly promoted employees struggle to master their new jobs, they will be continually pulled back into their former jobs to explain special nuances and procedures that are unknown to the people filling those positions.

Market globalization

Organizations of all types and industries are finding international expansion the answer to sustaining profitability and growth. With global access facilitated by websites and Internet-based communication technologies, former regional entities discover that their products and services are attractive to customers in the farthest reaches of the earth. "Local" has taken on a new meaning in this global economy. New international players quickly learn of the complexity of doing business with diverse cultures and regulations. Competing in broader markets creates a demand for new knowledge that will stress the resources of the most vibrant organization. Simply translating your sales brochures and employee guides into new languages won't address the challenge of doing business in new cultures and countries.

Technology advances

No one can debate the dependence businesses place on the use of technology for daily operations. Accompanying this technology reliance comes the demand to remain current, resulting in an insatiable demand for new knowledge. Technology, both hardware and software, must be updated regularly and often replaced. Skills, once highly valued just a few years earlier, become obsolete with new product announcements. Coupled with other knowledge loss factors, technology change widens the knowledge gap further.

Business complexity

A common lament heard from the corporate executive suite is the growing complexity of doing business. Some blame increased government and industry regulations and others point to the fear of litigation. No matter the source, most business leaders agree that it is harder to do business today than at any time in their memory. Technology becomes the only way to face this challenge, but that brings with it additional opportunities for a widening knowledge gap.

Knowledge opportunities

Much of the justification for a new knowledge management focus begins with the desire to protect vital intellectual assets from erosion, however, that is only part of the reason for making KM a priority. Equally important, KM is about finding ways to use your organization's knowledge more effectively. When blended with sound technology and an accommodating culture, the true value of KM will be found in how knowledge is more easily shared across organizational boundaries. Expertise held in one department or location can be readily shared with employees in any location.

Global knowledge sharing provides a more consistent response to all of your clients and dramatically reduces the amount of effort it takes to meet client needs. Instead of each district region or even country redundantly creating processes, procedures, and templates to respond to internal and external requests, a common framework can be established for all to follow. By engaging the knowledge experts from across your organization, you will also see this knowledge used more broadly and at a much faster rate.

Investing in innovations that can spawn new products and services is far more valuable to your organization than the redundant creation of fragmented organizational knowledge. When proven knowledge assets are included with organizational planning activities, your total business intelligence grows and opens new market opportunities.

Capturing and sharing the valuable knowledge of your organization gives you a competitive advantage over organizations who fail to do so. It increases your "win" potential. Knowledge management offers strategies to save unnecessary expenditures and find ways to generate new revenue streams.

Return on investment

Implementing knowledge management represents a significant investment for any organization. It costs time and money. It uses resources that could be putting out fires somewhere else in your world. Prudent corporate leaders will attempt to justify this expense by examining the potential for financial return. To arrive at a valid return on investment result, consider the following.

1. *Establish consensus:* It is critical that the decision makers of your organization have a common understanding of the basic realities of knowledge, organizational knowledge, and knowledge management. This book provides you with the information you need to lead this discussion. Failure to achieve agreement around these concepts will lead to false expectations and confusion.
2. *Align organizational KM efforts:* As you pursue a knowledge management strategy, you will discover many valid processes that directly support effective KM. Later in this book, I describe creating a *knowledge management vision statement* (Chapter 10) that helps you align isolated efforts with senior executive expectations for KM. This set of documented organizational goals provides a global roadmap of what is currently available in your organization and is the springboard to new directions and individual contributions.

3. *Start small, but critical:* Many organizations have launched a KM initiative for the total enterprise with their primary focus on records retention. As I describe later, these *enterprise content management* (ECM) solutions sound promising when proposed by a repository product vendor. Who wouldn't want all their records accessible electronically in one location? But the task of making this happen is so large and requires such detail that corporate patience soon wears thin. In the deep dive for minutia, we fail to establish a context for these records. The key to implementing KM is identifying segments of the organization that contain knowledge vital to your operation, knowledge that is in peril. Addressing this knowledge gap will provide tangible value. This book helps you create knowledge inventories, one knowledge domain at a time, following a stable repeatable process. As you apply this strategy to other business domains, you may integrate the results over time. You should also use knowledge artifacts found in active projects to fuel a KM strategy. Conduct a realistic assessment of the value to your organization if this source of knowledge is tapped for even one year.
4. *Capture metrics:* Keep track of the effort and costs associated with early KM activities. You should be able to refine those numbers into formal metrics that can be used to forecast future activities. By tracking this information, you will also be able to supply more credible numbers on the true cost of your KM implementation. Once you have something in place, you will want to capture the actual usage of the retained knowledge. The key to capturing these data will be a *knowledge portal*. This single point of entry to your knowledge assets will provide usage statistics that allow you to analyze the resources accessed to accomplish financial gain in your organization. When you accomplish a major organizational success, allocate a portion of those gains to the KM effort that helped bring it to reality.
5. *Display tangible results:* It is critical that your KM implementation not get lost in the weeds of granularity. Everyone—management and your KM implementation team—will want to see tangible results from your efforts. When knowledge management efforts are applied to a specific subset of your organization, initial results will be available within weeks, not months or years. These by-products will certify the value of the process to the organization and provide the structure for the more granular activities. Local success may then be replicated across your projects and departments.
6. *Engage management:* Although knowledge loss and replacement have implications for every employee, management must advance the vision for any KM activities. The beliefs and behaviors of your organization's leadership will create a culture that either encourages or discourages knowledge sharing and reuse. Organizationally

charged with making sure your enterprise runs smoothly, all levels of management should share in the total vision for KM. Use the knowledge management vision statement to help your leaders understand the potential for knowledge integration. Select an initial business domain based on having management that is predisposed to support this effort and make it successful. When creating the early components of the knowledge inventory, make the senior executives of that area active participants in the process. Nothing gains ownership more than direct participation in the capture of knowledge assets vital to your enterprise.

7. *Recognize and attribute success:* This is more than recognizing the personal efforts required to implement KM. It is associating portions of future financial gains to the effort required to make the knowledge available. When efficiencies are achieved by reusing existing knowledge, recognize it on the spot and attribute portions of the monetary gain to the KM effort. When you achieve internal or external success based on using established knowledge assets, allocate some of the benefit to your KM investment. When new knowledge is created and captured for future use, identify the value of this asset. Work with your human relations department to identify the potential costs of lost knowledge. Recognizing the value of your KM investment will encourage additional investments in needed areas.

A call to action

Perhaps the value of organizational knowledge is something you have contemplated for some time. Or maybe you are just coming to terms with the breadth of the situation. In either case, this is a time to take action. As described in the balance of this book, it is past time to get serious about the discipline of knowledge management and how you can use it to manage your intellectual assets. In these pages are strategies to help change the mindset of your organization to capitalize on knowledge work and knowledge workers. This will enable you to understand and accommodate the needs of knowledge workers so you can be more successful retaining and transferring what they know before they are gone. I also build the case for organizational knowledge as a true asset, perhaps your greatest asset. And we examine how you can create incentives to reward people for transferring what they know to your next generation of workers.

How you respond to these new realities may well determine the survivability of your firm. The time to act is now. If you wait until all of the conditions presented here are fully realized, it will be too late to take proactive steps.

When my daughters were young, we enjoyed a vacation on the island of Maui. I was holding the hands of my youngest daughter, Casey, while she bounced innocently in the waves. Two rock walls that framed our beach exaggerated small waves. It was an idyllic setting, jumping and playing as the waves rolled in. Alerted by the screams of people around me, I turned to face a wall of water headed directly toward us. A rogue wave had entered this formation and was becoming increasingly formidable by the second. There was no time to exit the water or even warn Casey before the water was upon us. All I could do was lift her out of the water and hold her as tightly as possible. The water swept me off my feet. I felt like I was inside a washing machine on the spin cycle. Strong currents pulled at my daughter, threatening my grip on her. I wasn't actually sure I still had her in my arms. I wondered if Casey had been able to get a breath of air or where this wave might take us. Within seconds (it seemed like days), we were deposited some distance up on the beach. Casey squirmed out of my grasp, stood over me and declared, "That was fun, can we do it again?"

Reacting and surviving is a good thing, in the ocean or with your organization. Being proactive is even more valuable. Seeing danger before it comes allows you not just to survive, but to thrive in a changing world. Are you seeing this tsunami heading your way or do you just plan to hold on to personal valuables and hope to ride out the knowledge gap? You can see this wave of change coming. Some of the implications are ominous. Is your organization ready? If not, hold on and hope for survival. Hope isn't much of a strategy.

chapter three

KM beliefs

> Live your beliefs and you can turn the world around.
>
> **Henry David Thoreau**

Before you plunge into the depths of this chapter and the next, it is important to first understand why this content is so significant. Over the next few pages, I describe a specific set of beliefs and formal methods that make knowledge management possible. These *knowledge management beliefs and processes*, or KMBPs, are what give substance to the KM discussion in your organization and will keep the topic from being dismissed as irrelevant or academic. You will be able to use these observations to help people see that they are already performing many of the elements that support KM. Without this framework, however, I find that people exert great efforts doing good things but having them focused only on local benefit.

As stated in Chapter 1, I define *knowledge management* as applying formally defined, repeatable processes that facilitate knowledge discovery, capture, organization, use, transfer, and retention to a specific domain of intellectual assets. The key to this definition is understanding and integrating the KMBPs. By examining a defined collection of KM beliefs and processes, you will be able to (1) establish a common understanding of knowledge management within your organization, (2) conduct a formal assessment to determine current organizational realities, and (3) establish tangible goals and strategies to implement KM in your organization. You will likely find that localized efforts exist within your organization that map to these KMBPs. The challenge will be to convert these local efforts into a "formal, integrated" approach.

Distinguishing "beliefs" from "processes" is sometimes difficult as they tend to overlap. In general, *processes* are a collection of activities that can be documented in formal methodologies or procedures. Processes are methods, strategies, and tactics that are defined, repeated, and taught. Some will be absolute formulas whereas others provide general guidelines; KM processes are what we "should" be doing. Chapter 4 describes a set of KM processes that greatly benefit effective knowledge management implementation.

Beliefs, on the other hand, establish the ground rules for organizational behavior. They define your culture. Behavior reveals what an organization truly values, especially when faced with difficult choices.

Behaviors are guided by stated beliefs that must be communicated consistently from senior executives to junior staffers. The intent of stated beliefs is to inspire specific daily actions that are consistent with the values of senior management. When an organization fails to establish a clear set of beliefs or allows the message to be compromised, people will become confused about their behavior and cynicism sets in.

Beliefs and processes are tightly intertwined and interdependent. Processes provide the structured actions that reinforce the beliefs. Beliefs establish the common assumptions that make the processes practical. What we believe explains *why* we exhibit certain behaviors and processes describe *how*. KM beliefs set the environment for the formal methods found in the KM processes.

A senior director at a large international corporation was required to provide annual ethics training to his staff. One year, he posed a question about accepting sports tickets from a vendor. His staff displayed their understanding of the corporate beliefs by quoting the value statement that forbade accepting gifts of significant value from external parties, especially vendors. To their shock, the director explained that this was a real scenario and he had enjoyed watching a postseason contest from the vendor's stadium suite. An incredulous member of his staff asked how he had justified this decision. The director stated that he had informed his superior of the decision to take the tickets and, although his boss had expressed concerns, the communication made the decision acceptable.

I'm not sure what this director thought he was accomplishing with his training exercise but I know what resulted. First, his staff realized that corporate beliefs and policy meant little to their boss. Second, should they wish to violate these ideals, they need only inform him. And lastly, they also lost all confidence in their boss' superior. The morale of that organization was devastated. Interestingly, this same director had publicly (and falsely) accused a former member of his staff of taking a vendor gift a year earlier.

Of all the topics covered in this book, none are more important than making sure you have clearly communicated a set of core beliefs that will enable a KM environment for your organization. No other issue has proven to be more beneficial or problematic. Corporate culture will always reign supreme over defined processes. You will find spotty success implementing the formal methods needed to support a consistent KM environment if your organizational beliefs are discounted by situational ethics.

There are many organizational beliefs that affect a knowledge management environment, however, the most significant address: (1) knowledge sharing and reuse, (2) organizational learning, (3) use of best practices, and (4) communities of practice.

Knowledge sharing and reuse

> *Belief: Sharing personal knowledge and using established knowledge increases the value of an employee.*

The most consistent questions I've heard when discussing knowledge management success and failure relate to encouraging staff members to share and reuse existing knowledge. If you have any expectations of your people making their individual knowledge available to others in the organization, they must be absolutely convinced that sharing their knowledge makes them *more* valuable to the organization, not less. The same is true when trying to encourage your people to use established knowledge instead of creating it anew.

It is easy for executives to make statements to this effect, but actions speak far louder than those words. How many times have you heard that "Our people are our most valuable asset" only to see actions that indicate otherwise? This belief cannot be compromised. In fact, the reverse of this belief is also true. The unwillingness of an employee to share his or her knowledge should make him or her less valuable to an organization. Efforts should be made to replace those skills so that the organization does not become overly dependent on knowledge hoarders.

Your staff needs to see tangible evidence that knowledge sharing is important. Most people focus on financial rewards but professionals are also motivated by public respect or recognition. Some find incentives in just helping others whereas many are willing to share what they know in exchange for some new knowledge they are seeking. The key is finding what motivates your staff and adopting formal methods that reinforce the value of knowledge sharing and reuse.

One proven way to make this work includes establishing individual knowledge objectives with every employee. Each person needs to recognize his contribution to the organization's knowledge pool and the importance of reusing existing knowledge. Chapter 10 explains how to set and validate these knowledge goals during the annual employee review process. To make these personal knowledge goals effective, it is important to first set an organizational vision for knowledge management. Chapter 10 also explains how to use the KM beliefs and practices to create a knowledge management vision statement for your organization. This will enable individual staff members to align their contributions with the organizational intent.

Although not as popular as a reward for sharing or using knowledge, there should also be consequences when these behaviors are ignored or trivialized. This could range from requiring an employee to attend additional education on the behavior, receiving personal coaching, or being excluded from participating in financial incentives.

In some instances, little can be done to make this work. There are too many organizations (not yours, I hope) who have demonstrated a repeated disrespect for the knowledge of their people. They have made it clear that people are a disposable resource with little consideration for the actual value a person brings to the organization. When trust cannot be established between management and staff, the damage is difficult, if not impossible, to recover.

Learning organization

> Belief: Knowledge increases with a commitment to organizational and individual learning.

There are very few organizations that would publically state they are not interested in being a learning organization. Imagine finding that on a recruitment poster. Just how interested would you be in working for an enterprise that chooses to ignore learning opportunities? In actual practice, however, many organizations do just that. The key is understanding how organizations learn.

Organizations do not learn on their own. Organizations are a structure comprised of people. Organizational learning is enabled by individual learning. When the individuals within an organization gain new knowledge, the organization gains new knowledge. Organizational learning is taking advantage of every opportunity to encourage individual learning. It is this individual learning that you want your people to share with the rest of the organization.

Again, the proof is in the actions of an organization. Here is a quick test. Does your organization support a rigorous *lessons learned* process? Lessons learned activities are a staple of both project management and business operations. Using a formal focus group from a significant number of companies, my research reveals that very few organizations regularly conduct project post-mortem sessions and even fewer distribute the results.

Lessons learned allow individuals to share their experiences, good and bad. Now reflect back on the knowledge sharing and reuse belief. How many people do you think are willing to share the results of a bad experience if they thought this act could get them fired or demoted?

Here is another question. The moment your organization or industry encounters a financial challenge, what happens to your training budget? If your organization is like most, training dollars are reduced or elimi-

Chapter three: KM beliefs

nated. By cutting individual learning, you are also limiting the options for organizational learning.

Training comes in many forms, from vendor workshops to internal programs or even active weekly reader programs. Unfortunately, training is commonly viewed as an optional employee perk. If that is how your organization perceives training, you are probably already wasting your money. Your people have likely figured out that training isn't intended to change anything. It is a ploy to keep them motivated with some paid time away from the office or a little intellectual stimulation.

Training should target specific behavior changes or new needed capabilities. This is easily accomplished when you clearly define the beliefs and processes needed to support an active KM environment. You identify the behavior or skillset desired and focus your training around those expectations. Selecting training should not reside entirely at the employee's discretion. Although you want their input in the process, career counselors should participate by mapping an employee's career goals to what the organization finds valuable. If the employee wants to pursue an area of interest that does not directly support the beliefs and processes identified in the organization's knowledge management vision statement, fine, but that is called a hobby! Unfortunately, without any clear directions from their organization, many employees are left to imagine what skills and beliefs might be valued and build career plans from speculation.

Formal processes should be in place to reinforce the training result along with a support structure that provides peer support. Most important, people should be encouraged to apply new skills while their new knowledge is fresh. This gives the staff the opportunity to refine their new understanding and have guidance from more experienced internal resources.

When you turn off the pipeline for new individual knowledge development, you interrupt the flow of organizational learning. Declare a commitment to individual and organizational learning. Then establish the knowledge management processes that help capture and transfer new knowledge.

Best practices

> *Belief: Proven best practices should be promoted and used.*

Of all the KM beliefs, support for using repeatable best practices may be the most significant. Best practices range from formal methodologies, proven techniques, strategies, tactics, or business processes identified within your enterprise. To obtain knowledge and skills in selected best practices, organizations often send staff members to specialized training or encourage them to participate in professional societies. Upon return to the office, however, these employees find there are scant options to use their newly learned skill. They are told to wait for a future opportunity when there is more time to apply the new knowledge. Change seems futile and the enthusiasm over this new skill quickly dissipates. This is a lost opportunity for individual and organizational learning.

Even more damaging is when an organization undertakes some special quality certification process that is intended to validate an organization's commitment to doing things the right way. Only once the ink is dry on the paperwork, the organization reverts to former habits. Although a lost learning opportunity, this action erodes employee confidence in the integrity of their organization.

Leaders can no longer survive organizational doubletalk. You will never implement the disciplines needed for knowledge management unless the total organization has committed to using legitimate best practices. Implement processes that formally endorse the use of best practices and recognize major advancements made by your people. Encourage people across the organization to nominate best practices that demonstrate innovative tactics and deliver successful results. Ask the practitioners to document what they did and store these suggested ideas in a highly visible and available repository. Provide some form of recognition or award for the monthly best practice and communicate the top three candidates to the total staff. You might even offer the selection of a finalist to a group vote. This practice will not only inspire people to contribute innovations, but also encourage others to build on proven best practices.

A company I consulted with had a team who created an automated tool to calculate very accurate estimates for their work. It was based on a carefully constructed decomposition of the work activities along with the typical times required to accomplish the work. To this, they also baked in factors for the complexity of a specific assignment along with the skills of the individuals who would be doing the work. It was a fabulous implementation of metric-based estimating. The estimating tool received an award by a quality organization at a national convention. It was not until later, accompanied by some embarrassment, that the team's management discovered what the team had achieved. When the dust settled, this estimating tool spawned three additional similar products and processes that were tailored to other environments. It was also the first in what became a long list of best practices formally recognized and publicized by the organization. That is the point of formally recognizing best practices in an organization. The hope is that by sharing this knowledge, others will be inspired to build on the idea or create something new.

Communities of practice

> *Belief: Communities of practice enable process improvements.*

Communities of practice are not new as the idea has circulated in many organizations under various names. Communities of practice, or CoPs, are comprised of peers who have some expertise, industry, or methodology in common. Participation is usually voluntary and each individual may join as many CoPs as they have time and interest. Professional societies are a CoP with a formal structure. Most internal CoPs are simply coworkers who agree to meet on a regular basis and take on challenges specific to their common interest.

When applied to KM processes, communities of practice are ideally suited to help identify, validate, and propagate the use of formal methods. They may also recommend templates, metadata, naming standards, and knowledge structures. CoPs can serve as advocates for knowledge transfer and use, helping to identify and resolve any discovered roadblocks.

Every best practice identified for use within your organization could be a candidate for a CoP. Active practitioners are best equipped to identify valid methods and any improvements or modifications needed to make the associated techniques more useful to your organization. They can recommend training that clearly communicates the finer points of using a best practice.

So, how many strategies has your organization pursued that sounded promising initially only to learn they don't deliver? Communities of practice will help preserve and spread knowledge of various topics across your organization. That way, different groups and individuals won't fall into the same misconceptions.

Although management should not mandate the formation of or participation in a CoP, they can encourage it. The most important thing is to trust the CoPs to make the right decisions. They should reinforce the belief that their employees are using their time in a way that will bring benefit and efficiency to the organization. They must trust their employees to represent all of the organization when making decisions within a CoP. If your staff does not believe they are empowered to make the decisions needed to promote their skill area, they will likely lose motivation and interest.

Encourage all members of your staff to participate in one or more communities of practice. Make it part of their personal development. What a wonderful opportunity for your new employees to learn from the more experienced members of your staff.

Conclusion

Defining and supporting a set of beliefs that support knowledge management is a critical prerequisite to implementing KM processes. Here are some action items to consider.

1. *Articulate KM beliefs:* Using the beliefs statements in this chapter, work with your senior executives to craft something similar. Include these statements in an announcement to the total staff of the intentions to implement formal methods that enable knowledge sharing and reuse. Encourage your senior management to personally deliver this message. With the array of communication technologies available today, there is little reason simply to send out another uninspiring memo. Look your people in the eye and let them know you believe what you are saying. Don't be surprised if there is some resistance. Many people carry the scars of previous "management promises." The best way to prove you mean what you say is to support specific processes that will make the beliefs a reality.
2. *Encourage specific behaviors:* Implementing knowledge management represents significant behavior changes for many of your people. Basing KM on specific Knowledge Management Beliefs and Processes (KMBPs) allows you to identify beliefs that might limit individual performance. It might be helpful to conduct a short survey to determine the mindset of your people. That will help you create programs to achieve the desired behaviors.
3. *Create educational programs:* Knowing what behaviors you hope to see, you can create curricula to guide the training for various groups of professionals. There is a variety of knowledge transfer mechanisms (see Chapter 7) that facilitate this process. Work with the relevant communities of practice to select the actual training courses that best fit your organization. Make sure to include educational support for behaviors that directly support Knowledge Management.
4. *Articulate KM vision and goals:* By clearly defining the KMBPs your organization supports, you will create a knowledge vision that all can understand. You can then list the current contributions of various departments and show how personal knowledge goals fit into the mix. In Chapter 10, I propose creating a knowledge management vision statement that accomplishes just this purpose. Without a formal set of KMBPs, a knowledge vision will be little more than flowery words.

There is also a negative impact when an organization fails to define a set of KM beliefs and processes. KM will be left to unguided individualistic interpretations. At best, you will see inconsistent unrepeatable results.

chapter four

KM processes

> Knowledge Management is an interdisciplinary field that deals with all aspects of knowledge processes including knowledge creation, discovery, acquisition, sharing, transfer, retention and organization. It encompasses technology, people and organizational practices.
>
> **Suliman Hawamdeh**

In the previous chapter, I recommended defining knowledge management through a specific set of knowledge management beliefs and processes, or KMBPs. I presented four very important belief statements that would set the environment for an effective KM implementation. Articulating these beliefs is an important step, however, it is now time to support those beliefs with action. Applying KM processes proves that an organization believes what it says and is willing to live what it believes.

KM processes are the formal tactics, methods, procedures, and strategies adopted by your organization that directly support your KM vision. To support the KM beliefs, I recommend grouping these methods around six core knowledge processes: discovery, capture, organization, use, transfer, and retention. Each has very tangible elements. By studying each of these KM processes, you will be able to fit current organizational capabilities and future efforts into an integrated framework. It will also help you find redundancies to consolidate and gaps that need to be filled.

Knowledge discovery

Most, if not all, formal disciplines encourage strategies for discovering knowledge. When you meet with your physician for a physical, the clinical staff will check your vital signs and ask a series of questions. The doctor then looks over recent lab results and your medical history. The physical exam includes additional questions. Your answers to questions generate new questions. All of this flows into an intentional act of discovery.

Prior to designing a new structure, architects meet with clients to understand their needs for a building, large or small. Pipeline engineers study terrain, weather, and applicable regulations prior to building new pipe networks. Attorneys conduct discovery sessions prior to launching legal action. Business analysts perform needs analysis and requirements gathering with end users to understand core processes prior to selecting a technology-based solution. Again, all of these are candidates for formal knowledge discovery methods.

Discovery methods come in many forms. Some help investigate problems, others uncover solutions and still others test and validate end results. Anchoring your KM implementation with discovery methods contributes to the quality of the collected knowledge. Instead of having each practitioner create his or her own approach to discovery, or ignore the process entirely, formal discovery methods increase the potential of collecting meaningful usable knowledge. Applying repeatable discovery processes significantly improves the quality of any resulting outcomes. Your practitioners are able to invest in learning the nuances of a formal method, including how complementary methods may be integrated to yield an even higher-quality result. All of this is highly unlikely when people just "wing it." Formal methods for knowledge discovery allow your practitioners to enhance their personal skills faster and then add their own insights.

Knowledge capture

Knowledge capture picks up where knowledge discovery leaves off as it provides the vehicle for recording the uncovered knowledge. Capture is most effective when defined templates are used to record discovered knowledge. Templates encourage consistency in both content and format. This makes it easier to document significant knowledge and then find it later. A critical component to capturing knowledge is assigning, or *tagging*, descriptive information, *metadata*, to content. Often described as "data about data," metadata is simply a list of terms or descriptors for important content that describes a knowledge artifact. In elementary school, your teacher explained the role of adjectives to describe a noun. Metadata does the same thing. If you work with any office software products, take a look at the "properties" area. There you will find places to add the name of the author, the title of the work, and the subject being discussed along with keywords and comments. All of these are metadata.

Search technologies rely on metadata tags to help knowledge users locate the right content. In addition to establishing templates that are specific to types of knowledge content, recommended and required metadata should also be identified. Knowledge capture should include defining templates for significant organizational knowledge along with required and recommended metadata.

Creating knowledge content templates and metadata should be the work of peer groups or communities of practice who share an interest in the capture and use of a common knowledge type. Engineers should define the standards for engineers, medical professionals for medical professionals, and so on. These professional groups are a highly valuable source for building useful, considered templates and they can help make sure any recommended standards are accepted in the organization. These are the people who create and use this captured knowledge.

The options for templates are endless, but many may be found by examining industry recommendations for a specific discovery process. Sample templates may then be customized to the unique needs of your organization.

Your practitioners will gain additional benefit from the use of templates when they are not forced to invent them. When every practitioner captures knowledge in her own way, not only has your organization recreated the Tower of Babel, but you have made the resultant information unusable by anyone else in the organization. By using standard templates, your people will also be able to find multiple good examples created by others to help them along.

Over the years, I have formalized numerous discovery methods and designed hundreds of templates. When I work with a company implementing one or more of these discovery methods, the first step is to examine, refine, and even rename associated templates so they fit well in their organization.

Knowledge organization

To facilitate the capture of records and documents, most organizations have invested in a variety of repository products. Expectations were, when coupled with powerful search engines, these repository tools would provide fast and easy access to stored knowledge. That promise is proving elusive.

During a focus group project with representatives of several major corporations, I posed a series of questions about their repository practices. All the participants responded with real-life concerns. Several referred to their repositories as "dumping grounds" for inconsistent unsearchable content. One study participant stated, "You have to know exactly where something is in order to find it. That has led to people copying (or recreating) the same documents to multiple locations."

One engineering manager with over seven years of repository experience attributed the problem to a lack of consistent structure, minimal use of metadata to tag content, counterintuitive file names, and a failure to see the repository product implemented across the total organization.

In addition to recommending template and metadata standards for distinct practitioner groups in your organization, communities of practice should also recommend folder structures and file name standards that will make content easier to find. For example, look into the project management practices in your organization. Ask the following questions:

- What categories of knowledge are typically maintained to manage each project?
- Are consistent templates for this information used by all of the project managers in the organization?
- Are there consistent file names for these documents?
- Is standard metadata used to tag each document?
- Are a standard set of folders used to hold the information?
- Is the same repository product used to hold this content?

I've asked these questions of project managers for years and the most common response is embarrassed silence. And we wonder why we are unable to find important knowledge, let alone share it?

When trying to find better ways to facilitate the organization of knowledge, help is available. Inquire of the professional societies that support your industry. Many of these entities have recognized the challenges facing their membership and have created specialized committees to help with recommendations. Another source may be as close as your local library. Librarians have the training, standards, and personal discipline needed to create organizational structures for content. Library collections are organized using formal taxonomies for both content and metadata ensuring specific works may be located when needed. Many of the principles used to organize content at a library may be used in your organization.

Although the ability to organize content is critical for effective repository use, it is equally significant for group knowledge on shared hard drives. And for the record, I recommend that projects set up a structure in a project portal using five primary folders: (1) project charters, (2) project plans, (3) status reports, (4) issues, and (5) change requests. They may then add one or more folders to hold any other miscellaneous information needed for a specific project.

Knowledge use

The ultimate objective of any KM implementation is to facilitate the reuse of existing organizational knowledge to improve performance and reduce costs. It is vital that your organization explore formal methods and strategies that encourage and facilitate the widespread use of discovered, captured, organized, and transferred knowledge.

The benefits should be obvious and very attractive to your senior leadership. Knowledge transfer and reuse not only breaks important know-how out of isolated areas of your enterprise, but it becomes the most significant means to see that knowledge increase. Nothing expands knowledge like real-life application. The moment your people begin using established knowledge, however it was attained, they will gain new insights and innovations. The benefits of knowledge use should then be captured and shared with others. Communities of practice provide a natural forum to exchange such new ideas.

Setting personal egos aside, individuals in your organization have a great deal to gain from utilizing established organizational knowledge. Instead of following an uncertain "trial and error" process, they are able to capitalize on the experiences of others in similar situations. They will have greater confidence in the effectiveness of this proven knowledge. As a creature of iteration, these professionals will find it easier to contribute their experiences. They can routinely experience Isaac Newton's observations that "If I have seen further, it is by standing on the shoulders of giants."

If you expect knowledge use to occur within your organization, it is vital that you create the infrastructure and incentives to make this behavior a natural part of the business day. Creating a technology-based knowledge portal is a proven method for encouraging the use of knowledge. It enables people to locate the content they seek. The portal should provide users with effective search options and alert them to new captured knowledge. It should also link them directly to internal and external sources of additional help when needed. To make this natural, consider incorporating this portal into the ultimate home page for your staff. Knowledge portals are an effective mechanism for making knowledge use a natural business process. There is more about knowledge portals in Chapter 10.

Equally important to having a meaningful knowledge portal is making sure your people can actually find the needed existing knowledge with minimal effort. If your knowledge organization strategies are weak and the search engines lack sophistication, people will grow weary of rummaging through your corporate attic for what they need. They are more likely to surrender and just re-create the knowledge. When you search the Internet for content and the search engine returns several thousand possible hits, how many pages of possible links do you review? The keys to making this search more fruitful lies in the templates you used to capture the knowledge, the metadata that were used to tag the content, and how the knowledge was organized in a repository.

As is the case with sharing personal knowledge, it will also be important to provide proper incentives for your professionals to use it later. Consider setting personal and departmental goals for using established knowledge. You may even be able to track this usage by examining patterns from your knowledge portal.

Make certain that the content in your portal remains useful. This is accomplished by assigning content stewards to each cluster of organizational knowledge.

Knowledge transfer

A primary driver for a knowledge management program is to ensure the transfer of knowledge from one generation or location of employees to another. Formal knowledge transfer methods exist to support documentation, training, apprenticeships, mentoring, cross-training, and communications. Each of these *knowledge transfer mechanisms* is examined in greater detail when we review the knowledge inventory in Chapter 7. Your challenge is to discover, capture, and organize knowledge that is significant to the organization and then specify one or more ways that knowledge should be passed along.

The ability to consistently transfer established knowledge across your organization should be viewed as making your enterprise more efficient while helping you achieve a competitive edge. Instead of vital knowledge being located in isolated silos of your enterprise, effective transfer of knowledge assets allows you to offer consistent services across the organization. Spreading the knowledge around also relieves the pressure placed on specialized subject matter experts who become overloaded responding to simultaneous requests.

Formal methods for knowledge transfer will ensure the continuity of knowledge within your organization. Failure to employ deliberate formal methods for each vital knowledge asset leaves transfer to localized inconsistent efforts. It will also force you to deal with isolated pockets of expertise.

Knowledge retention

The ultimate driver for KM is to ensure the retention of vital organizational knowledge over time. It is this vision that justifies the needed investment in methods and strategies to support knowledge discovery, capture, organization, use, and transfer.

Knowledge retention is rooted in a component of enterprise content management (ECM) called document retention. Document retention is primarily concerned with keeping important physical records to answer common inquiries or respond to some type of audit or legal request. Just as knowledge management has a larger scope than ECM, the same is true for knowledge retention.

Knowledge retention is focused on several key issues:

1. *What retained content represents a significant knowledge asset?* Not all knowledge is equally valuable. Some knowledge actually obscures important knowledge.
2. *Who are the organization's best experts on each knowledge asset?* How can these experts be contacted when there are questions? Providing a knowledge "yellow pages" will facilitate a fruitful search process.
3. *How will each knowledge asset be transferred across boundaries and generations?* Specific, deliberate transfer strategies will help preserve important knowledge.
4. *If the knowledge asset is captured in a physical incarnation, where is it located?* This may include who has it or an actual link to the location of the knowledge.
5. *Of the valuable knowledge assets, which are the most significant to the operation of the organization?* Clearly, greater emphasis should be placed on knowledge that has long-term value and must remain current through regular review.
6. *Who in the organization is responsible for the ongoing evaluation and refinement of each knowledge asset?* Unmaintained knowledge, like a poorly maintained vehicle, quickly loses value.
7. *When and how should content be archived or destroyed?* Don't keep content that is outdated or inaccurate.

Although I found a wide range of formal methods to support the other KM processes, knowledge retention methods seem to be primarily focused on document management. In response, I designed processes and templates to construct a two-level knowledge retention policy that answers the questions listed above and more. The knowledge retention policy, or KRP, is designed to declare the knowledge assets significant to an organization. Chapters 7 and 8 explain how you can create a Level 1 and Level 2 knowledge retention policy to ensure you are recognizing and transferring the right organizational knowledge.

Knowledge retention is the capstone for any KM implementation. It is quite feasible to find formal methods for knowledge discovery that lead to good templates and metadata for knowledge capture. A little formal structural work will provide the knowledge organization needed to create searchable content. Emphasizing knowledge transfer methods for vital knowledge and encouraging knowledge use ensures the spread and application of existing knowledge across the total organization. But it is knowledge retention strategies that provide your organization with a long-term approach for knowledge management.

Conclusion

Defining knowledge management using a concise set of processes provides your organization many benefits. It gets everyone on the same page and helps you make this a very tangible discussion. Consider the following to build on this potential:

1. *Document executive goals for each KM process:* In the earlier chapters, I described having your senior executives create a grand vision for KM in a knowledge management vision statement. The process should begin with value statements about each of the KM beliefs listed in Chapter 3. Now add to that initial effort by having your executives create goal statements for each of the KM processes. In Chapter 10, I provide you with the building blocks for defining meaningful goals that can be supported by measurable objectives.
2. *Compile an inventory of current KM processes:* Once your senior management has declared their expectations for each of the KM processes, conduct an assessment of the organizational units that are included in the KM initiative. Look for formal methods and technologies that directly support each KM process. The results may not be integrated across organizational boundaries, however, this review will reveal the valid efforts underway in your organization. This alone will provide reinforcement to practitioners that their efforts are worthwhile.
3. *Identify commonalities and redundancies:* An initial inventory of processes and technologies that support each KM process will also likely demonstrate overlapping and redundant activities. Don't assume this is something to eliminate. It is not uncommon for different departments to address the same KM process using different strategies. A more in-depth evaluation, however, may reveal opportunities to link efforts and consolidate funding.
4. *Propose future directions:* The need to capture and share knowledge is not new. One benefit of this inventory will be finding the KM visionaries who exist in your organization. Use this opportunity to discover what innovations they are planning as they move their unit forward. Again, you may discover numerous opportunities to join forces.
5. *Review results with management:* Once you have compiled the base information in the KM vision statement, present the results as a living product to your senior management. This will allow leadership to gain a more detailed understanding of what is being done across their organization and will help them know about future efforts that need their support. It will put departmental requests into a larger context. Make sure to keep the KM vision statement current as "future" strategies become reality.

6. *Create a KM community of practice:* As you compile the information for the KM vision statement, you will find people in your organization who share a passion for implementing knowledge sharing and reuse. Consider organizing these people into an informal community of practice to help guide the KM implementation. This group is the ideal place to maintain the KM vision statement.
7. *Align individual skills:* In Chapter 3, I recommended asking each of your staff members to create individual knowledge objectives during their annual assessment. There is a more comprehensive description of this process in Chapter 10. As I describe there, your people will be able to identify more meaningful knowledge contributions when they are guided by your executives' vision for each KM process and they understand the actual tactics currently used or planned.
8. *Implement at the project level:* Using the information described in your knowledge management vision statement, live the KM beliefs and apply the KM processes at the project level. This will validate the reality of the KMBPs and will show immediate results.

I have provided a template for this Knowledge Management vision statement in Appendix A. In Chapter 10, I distinguish between enterprise knowledge goals (EKGs) and supporting group knowledge objectives (GKOs). Understanding the KM processes described in this chapter is key to the development of this document.

Defining knowledge management with a distinct set of beliefs and processes will allow your total organization to grasp a common definition for this powerful emerging discipline. Instead of focusing only on repository products, it will help people understand a broader view for a KM initiative. With this perspective in place, it is now time to move on to the real meat of knowledge management by answering some very significant questions. These include (1) what is knowledge, (2) what is organizational knowledge, and (3) how do we manage organizational knowledge? Keep reading.

chapter five

Defining organizational knowledge

> Information anxiety is produced by the ever-widening gap between what we understand and what we think we need to understand. It is the black hole between data and knowledge, and it happens when information doesn't tell us what we want or need to know.
>
> **Richard Saul Wurman**

The need for a formal discipline to counter the growing knowledge gap is not new. In the late 1990s, visionary authors speculated about the emergence of knowledge management (KM) and potential solutions it might bring. Although this awakening was good, many of the proposed strategies provided limited tangible value. Some even caused confusion and ineffective solutions.

Even if you are comfortable with your understanding of KM (and that should be your first warning), you may find something new in this chapter. What makes this discussion most important is establishing a meaningful definition for *organizational knowledge*. Although many authors have recognized the existence and significance of this special brand of knowledge, few have actually defined it. The definition presented in this chapter provides the foundation for the strategies offered in the remaining chapters of this book.

But before we take on the unique characteristics of organizational knowledge, let's start in more familiar, but often misunderstood, territory: what is *knowledge*? *Webster's* defines knowledge as "the fact or condition of knowing something with familiarity gained from experience or association." To fully understand the implications of organizational knowledge, however, it is important that we examine the categories of knowledge and the distinctions among *explicit, implicit,* and *tacit* knowledge.

Knowledge categories

When you review knowledge management literature, you will find the discussion often centers around three common categories of knowledge: *individual, global,* and *organizational*. There is much debate over what each of these represents, however, their distinct existence is obvious.

Individual knowledge

The original form of knowing is *individual knowledge*. Long the discussion target of philosophers, educators, and scientists, this form of knowledge resides in the minds of people. It is based on a mix of formal education and personal experiences. Two people, exposed to the same education and real-life conditions, will come away with different forms of individual knowledge.

The thoughts and understandings of an individual are his or her own. This is especially perplexing to business organizations when employees leave and take their individual knowledge with them. Although there are some legal options to govern new innovations created "on the job," it is impossible to coerce an employee to divulge private knowledge or prevent the employee from thinking about it in the future. Such is the domain of spy novels and science fiction movies.

Global knowledge

Formal knowledge management was initially focused on global or encyclopedic knowledge. This is knowledge that is or should be available to anyone wishing to access it. This type of knowledge is found in textbooks, reference guides, public databases, government publications, industry standards, and journals. Many of these resources are free whereas others require purchase or some type of subscription.

Today, most associate global knowledge with the Internet and high-speed connectivity. It has, however, long been the focus of professional librarians. Organizing stacks in a library brought us the concepts of metadata to describe the contents of a particular work, *tagging* or associating metadata to a work allows us to find it later, and *taxonomies* provide organizational structure to a collection of works or of the *tags* used as metadata. Although these terms may be unfamiliar to you, they are common to your nearest librarian. And you can thank the professionals from this industry for many of the standards that rule the classification, organization, and searchability of content on the World Wide Web.

Recognizing, tagging, and storing a staggering number of works in a collection also gave us some of the early efforts for knowledge management. Most of these strategies centered on purchasing and implementing *document management systems.* These technology-intensive solutions allowed organizations to store their records with the plan to search for them later. As I describe in the Introduction to this book, this first generation of KM thought yielded terabytes of confusing unsearchable content.

Some of this frustration may be due to limiting our thinking to include only a library context. Librarians tend to value all content, no matter how obscure or controversial. When this principle is applied to businesses, however, the result is a colossal mass of content where valuable materials are often obscured by worthless products. We cannot abandon the standards offered by library sciences, however, organizations must clarify what content they actually need to retain for future use.

Organizational knowledge

That brings us to organizational knowledge. Although this category of knowledge borrows some characteristics of individual and general knowledge, it is also quite unique. I define organizational knowledge as a formal integrated application of knowledge beliefs and knowledge processes (KMBPs) to an organization's data, information, and decision making within a specific subset of an enterprise.

Based on the past two chapters, this definition should be more meaningful than when you first encountered it in Chapter 1. I explain the "data, information, and decision-making" significance in the next chapter.

Organizational knowledge is the knowing required by an enterprise to produce the products and services necessary to perform the work of the enterprise. Organizational knowledge is commonly an intellectual property that is wholly owned by the enterprise using vehicles such as copyrights, patents, and trademarks. Portions may have been obtained by way of purchase or a license, but significant amounts of organizational knowledge are the by-product of the thoughts and actions of the employees. Your employees obtain knowledge used by predecessors and add their own perspectives and understanding. They may also reference general knowledge or some collection of purchased knowledge. The outcome is an approach to doing business that is unique to your enterprise.

Many of the issues surrounding the ownership of organizational knowledge is determined by when, where, and how employees created the knowledge. When organizational knowledge is created during the job-related conduct of a paid employee, ownership of the results is usually held by the employer. When this becomes an issue, legal standards are used to resolve the matter. Complicating the notion of ownership is the distinction between explicit, implicit, and tacit knowledge.

Explicit, tacit, and implicit knowledge

For centuries, philosophers and educators have debated the reality and distinction between explicit and tacit knowledge. More recently, debates encompass the significance of implicit knowledge. Understanding organizational knowledge and implementing meaningful KM strategies make this one of the most important knowledge issues on your corporate landscape.

Explicit knowledge

Explicit knowledge is any form of knowledge that has been captured or recorded using a formal physical mechanism. This could include writing content on paper, entering it into a computer, making an audio or video recording or even etching it onto a stone tablet. Just documenting a mess doesn't make it knowledge. The term "explicit" implies something that is clearly revealed with little or no question as to the meaning. It should be an unambiguous expression that does not require interpretation. Common examples of explicit knowledge in organizations would be procedures, policies, job descriptions, reports, proposals, contracts, transactions, and checklists.

Organizations expend significant resources attempting to capture their explicit knowledge and then keeping it current. Refinement of explicit knowledge is seen every time a chemist proposes a new formulation of an existing product or an employee group recommends a new strategy for working with the customer. Much of formal education is oriented toward explaining explicit knowledge.

Look around you. What contribution have you brought to your organization to explain how something is done or to improve something started by someone else? Did you document it? Have you recorded it? If so, that is explicit knowledge.

Tacit knowledge

Tacit knowledge, on the other hand, is inexplicable; at least it starts out that way. It is a form of knowledge that exists but cannot be captured in a tangible manner. It is knowing with certainty about something that cannot be reduced to words or physical media. Sometimes considered as intuition or judgment, tacit knowledge may be the most important organizational knowledge in existence, and the most challenging to transfer. But it may also be where your organization is at greatest risk of loss when seasoned employees walk out the door. You may hand their documented procedures to replacement staff, but the new staff fails at the job because they don't have the understanding that lives between the lines of the explicit knowledge.

Chapter five: Defining organizational knowledge

The story goes that a company's operations were halted when a major piece of equipment simply stopped running. After frustrating attempts to restart the machinery, someone suggested bringing back "old Joe" who always knew how to make the equipment hum. The desperate owner promised Joe the moon to get things back up and running. Joe walked the factory floor, carefully examining the total layout. He then used a small hammer to give a sharp tap to the side of the equipment. Instantly, everything started working again. Employees cheered and the owner was thrilled, until Joe handed over an invoice with a total due of $1,000. The owner immediately protested, "A thousand bucks to hit the equipment with a hammer?" Joe asked for the invoice back and made a revision. It now read, "Striking the equipment with a hammer . . . $10. Knowing WHERE to strike the equipment with the hammer . . . $990."

Although this story is likely contrived, real-life examples surround you. Your organization is brimming with knowledge workers who can explain only part of what they do. There are elements of their knowledge, vital to your organization, that are simply inexplicable. Later in this book, we discuss how to transfer tacit knowledge. For now, it is critical that you orient your thinking to recognize and respect this most valuable form of organizational knowledge. And I leave it to the attorneys to determine who actually owns it, the organization or the individual.

Implicit knowledge

Many people, myself included, find there is actually a third form of knowledge. It is knowledge that is not formally captured, but it could be. In its undocumented state it may technically be tacit knowledge, but it doesn't need to remain that way because it can be defined. Converting implicit knowledge into explicit knowledge is often a matter of asking the right people the right questions and capturing the results. I believe this process is the most significant challenge and opportunity for third-generation knowledge management. More on this in the next chapter.

The problem with implicit knowledge is that it starts out being tacit. Following extensive evaluation and distillation, it gets documented, becoming explicit. So the classification of any knowledge as "implicit" is simply an after-the-fact observation. I believe it is valuable to consider this a distinct type of knowledge as it makes us more observant and willing to challenge the unknown basis for tacit knowledge.

Some implicit knowledge remains tacit for political reasons. In other cases, employees could document their knowledge but refuse to do so, viewing their knowledge as job security. Many organizations are guilty of promoting this belief when they penalize employees for sharing their knowledge. We read daily of organizations deciding to outsource or off-shore operations. They ask seasoned employees to help explain important processes and procedures to the new group, only to terminate their experienced staff when the transfer is complete. Not only does this create deep resentment, it sends a message to everyone else in the organization that knowledge workers are not respected. As a result, people hold on to their knowledge instead of sharing it with others. Implicit knowledge, or knowledge that could be made explicit, leaves with the former employee.

Does this reality doom subcontracting or outsourcing work? No, but it should raise a caution flag when making such a decision. If the work function is straightforward and clearly explicit, it is a candidate for outsourcing. If the knowledge needed to do the job requires undocumented experienced judgment, outsourcing will only make matters worse.

A major logistics company tried repeatedly to create a technology product to solve a critical scheduling problem. After two failed internal attempts, management elected to offshore the project. Years and millions of dollars later, the contractor delivered a product that was so flawed and incomplete that it was immediately abandoned. The foreign contractor had limited experience in the target industry. Due to time-zone differences and bad decisions, they had minimal access to the subject experts within the company. In other words, they couldn't even tap into the explicit knowledge of the organization and had even less access to tacit knowledge. But they were cheap. When the project failed for the third time, management wrung their hands in collective disbelief. They simply could not understand what had gone wrong.

Although some tacit knowledge will never be physically recorded, there are strategies organizations can follow to convert implicit knowledge to explicit knowledge. These include:

1. *Providing the opportunity and facilities for your employees to reflect on difficult problems and simply think.* This means giving them an environment that encourages problem-solving without interruption. This includes providing a place where people can work without the constant noise of daily work, including cell phones. When do you have your true breakthrough thoughts? Where are you able to be most productive? I'll bet it is when you are able to step outside the distractions that surround you. A physician told me he keeps a grease pen in his shower as that is the only time he is alone with his thoughts. I'll bet his wife loves that practice. I'll just keep a note pad next to my bed.
2. *Continually exposing people to new explicit knowledge.* Explicit knowledge may come in the form of classroom training, periodicals, articles, or even books such as this one. Provide your employees with the opportunity and incentives for life-long continuing education. While traversing some obscure idea, they may suddenly discover a new way to do things.

3. *Implement and observe a lessons learned process.* If your organization truly believes in quality, one characteristic to strive for is never to repeat the same mistake twice. Conversely, you should also look to repeat the processes that brought previous success. With the work of your organization spread across far-flung elements of the organizational chart, this will require a formal intentional act. The best way to achieve these ideals is to implement a lessons learned process that looks for both the good and the bad. It isn't focused on punitive acts but sees every event as a learning opportunity. To be useful, these observations must be shared across your organization.
4. *Encourage people to establish mentor/mentee relationships.* As the knowledge level grows in a mentee, they will begin to recognize and respect actions by a more senior person. That will lead to a greater understanding and maturing process. Mentors also gain value from the interaction as they respond to questions and are able to look at the situation through a fresh set of eyes.
5. *Apply improved interviewing and analysis methods.* Often, we are able to capture explicit knowledge that was considered tacit only because it was never collected from the right people. Encourage people to research their profession or discipline to discover better ways to know what questions to ask and how to follow up for the right information.

Knowledge characteristics

In addition to understanding the distinctiveness of organizational knowledge, it is also important to recognize some very unique asset characteristics associated with knowledge. Here are a few fun realities to consider:

1. *Knowledge is an intellectual asset.* While explicit knowledge takes on the form of a physical asset, the knowing associated with tacit and implicit knowledge resides in the mind of an individual.
2. *When knowledge is shared, there is no loss to the original.* When physical assets are shared, the "giver" must surrender something in the process. Not so with knowledge.
3. *Sharing knowledge often increases its value.* This characteristic is the by-product of interactions between both the person sharing the knowledge and the people receiving it.
4. *Knowledge can be shared endlessly.* Because there is no loss that results from sharing, knowledge may travel as many directions as there are vehicles to transport it.

5. *Once shared, knowledge assets cannot be recovered.* Although you can regain control of a physical asset, that is not true with knowledge assets, unless you've come up with one of those "flashy, blinky things" from *Men in Black* that wipes out memory.

Why bother?

Even if you didn't realize it prior to reading this chapter, capturing and managing your organization's knowledge is a nontrivial challenge. The question will be raised of why this is important. You will need to create your own priorities to justify the effort but here are some thoughts to get you started.

1. *Continuity*: The most obvious reason for capturing and transferring organizational knowledge is to achieve continuity of operations. Your organization was built on a reputation for performing a specific type of work or delivering a unique product. Failure to retain the knowledge needed to continue this capability results in lost customers and the associated revenue.
2. *Consistency*: A less obvious reason to recognize the existence of both explicit and tacit knowledge is to establish a consistent way of doing work in your organization. In all likelihood, you will find people in different parts of your organization doing similar types of work, but doing it in very different ways. Sometimes, these variations are based on individual skills and preferences. Many times, however, one approach may be far more effective than another. Unless these uncommon solutions to common problems are recognized, they will not be refined.
3. *Compliance*: Most organizations are required to comply with industry and government standards. Yet when individuals or distinct groups in your organization are left on their own to meet these requirements, they will do more than just add personal interpretations. They may also unintentionally place your organization in violation of these mandates.
4. *Standards*: The failure to capture knowledge and share it across the organization limits the use of peer-based communities of practice. Communities of practice are a vital component for setting performance standards and building on established knowledge.
5. *Quality*: One characteristic of quality is the absence of rework. Without a strategy for knowledge capture and sharing, your staff will never know about knowledge that has been created by another part of your organization. As a result, they will unnecessarily re-create procedures, templates, and outcomes. This also opens the

door for error or inefficiencies. Worse is when the same people must re-create their own work because they cannot find the earlier results.
6. *Lessons-learned*: When coupled with a formal lessons learned process, knowledge sharing helps your organization dodge the potholes discovered by another group. How often have you seen different groups in the same organization fail due to the same mistake? It happens far more often than we want to admit.
7. *Understanding*: The failure to share knowledge across an organization can also lead to false interpretations of the results. During the investigation into the Challenger disaster, NASA was hard pressed to believe that foam had caused the problem. They knew that foam had the tendency to break away from the external fuel tank. In fact, they referred to the phenomenon as "pop-corning" due to the small marks the foam pieces would make on the heat-absorbing tiles. The condition was so common, NASA engineers simply dismissed it as being "normal." Reports on the Challenger disaster referred to this as the "normalcy of deviance." They had the knowledge in front of them but failed to understand its full implications. How many deviations or convolutions has your organization allowed to become a normal act of doing business. Failure to expose these common but unusual behaviors on a larger level may well set your organization up for a major disaster.

Conclusion

Creating a plan to manage your organizational knowledge must begin by challenging core beliefs about knowledge in general. The failure to explore the distinctiveness of organizational knowledge or only recognizing explicit knowledge will dramatically undervalue your most important assets, your people.

chapter six

Recognizing organizational knowledge

> Where is the Life we have lost in living?
> Where is the wisdom we lost in knowledge?
> Where is the knowledge we lost in information?
>
> **"The Rock," T.S. Eliot**

In the previous chapter, I distinguished between tacit, implicit, and explicit knowledge and made the case for organizational knowledge. But how will you recognize *organizational knowledge* when you bump into it? I believe I can clarify the components of organizational knowledge by starting with a framework that is commonly used in KM writing, a framework that I believe helps and hurts. It is known as the WKID hierarchy.

Rather than being the call sign for a radio station east of the Mississippi, WKID is an acronym for *wisdom-knowledge-information-data.* This structure is intended to show how data becomes information and then information becomes knowledge with wisdom resulting from the three. If you have done any prior reading into KM, I'm sure you've come across this argument. Many authors go even further by adding elements such as judgment and intuition.

As you see, I like the distinctiveness of data from information, but that is where I get off the WKID train. I consider this structure highly susceptible to challenge as it is not the by-product of deep academic research but may have been inspired by the T.S. Eliot poem quoted above.

But that is not why I dislike the structure. I find it flawed because it is a circular illogical definition. If you were asked to define the word "girl," you would not reply with, "Well, it is a girl that . . . ." The word being defined or described cannot be used in the definition or description. So that knocks out the "K." I also challenge euphemistic terms such as "wisdom," "judgment," or "intuition" as meaningful contributors to understanding organizational knowledge. Although I concur that some people are wiser than others, some make better judgment calls, and still others seem to have a higher level of simple intuition, these are observations on behavior and most likely the result of their tacit knowledge. I also find it is totally possible to appear wise by simply remaining silent.

In place of WKID, I propose the DIDM framework of *data, information,* and *decision making.* Let me start with the two that have the greatest consensus among KM thinkers: data and information.

Data

Data represents the basic terms used in the operation of an organization. These terms should take the shape of formal definitions that are consistent and nonnegotiable across the total domain of an organization. At a most granular level, the term "flight number" must have the same meaning regardless if it is being associated with a passenger reservation, assigning gates, or planning catering. Yet one major airline discovered 17 distinct definitions for flight number within their enterprise. In a clinical setting, we discovered multiple interpretations for the word "stat." Some health providers used "stat" to communicate immediacy, whereas others meant "as soon as you have time." Still others indicated that it described a timeframe that included the end of the business day. This variation in interpretation may seem trivial, unless you are on the gurney waiting for a medication to relieve your pain—right now!

I'll bet you are already coming up with similar conflicts in your own experiences. Millions of them abound. I once had a debate with a young print shop employee over the term "collate." How do these inconsistencies happen? Simple. The terms were defined, not as a universal language within your organization, but within independent silos. Each part of the organization gave definitions that fit their perspective on the data. I once managed a project where our team spent over a week working with internal business experts to create a singular vocabulary that could be used by people in our Midwest and Northeast offices. The terms were not based on local jargon or strange accents, just different words for the same concept. The concept of data may also be applied to larger collections of facts such as this book, a song, a database, or your filing cabinet. For the most part, however, organizational data is quite granular.

Before an organization can learn to manage its knowledge, it must first create consistent definitions for the data it uses. Failure to do so results in overreliance on special interpretations, convolutions, and "work-arounds" to facilitate inaccurate data meanings. Without meaningful data, resulting information will be flawed and unreliable. Defining consistent meanings for the data component of your organization is exhaustive and exhausting. There are, however, multiple disciplines and best practices from a variety of industries that help you build thesauri, glossaries, data dictionaries, and data models. These methodologies have been a staple in the creation of automated business systems since the early 1980s. Integrated automation that spans an organization has become an instrument to force some level of consistency in data meanings.

Here are some characteristics that help you recognize the data assets within your organization. Data:

- Has a tangible name
- Has a single definition across the total organization
- Represents nonnegotiable meanings about your organization's business
- Is static; it simply exists
- May be a single element or a large compilation of static elements
- Is used to create information

Getting data right is a laborious process. It means inspecting every term used by your organization. Untrustworthy data meanings will automatically taint any resulting information. So, do you want your physician making a diagnosis of your condition on lab results that have no standard definition? Unfortunately, it happens.

Information

Information is comprised of data arranged into a deliberate structure. The most significant characteristics of information is that it: (1) implies movement, (2) must have a sender and receiver, (3) is considered valuable to both the receiver and sender, and (4) has a purpose.

Static data is at rest where information must be active. What distinguishes information from noise is the value placed on it by the receiver. As it moves from one location in an organization to another, it is generally expected by the receiver, it is valued by the receiver, and it has a legitimate purpose to the organization. Is everything you receive as "information" something you want and value? Heavens no! That is why we have spam filters and trash cans. How much of the physical mail you receive makes it through your door?

The composition of information may be recorded using the same methods used for data definitions, but more important are the paths information follows within the business environment. When models are created showing how information travels from sender to receiver, business experts are able to validate the legitimacy of the movement and the need for the information. Often, we find that information that was once valued is no longer needed or there is a redundant version of the information floating around. These now represent a negative value as they prevent productive work as you must continually check to see which one is "most" correct. Data and information that remain undefined force an organization to operate on tacit intuition and perception.

Some interesting characteristics of information are:

- It is trustworthy only if comprised of trustworthy data.
- Information moves between a sender and a receiver. If one of the two is missing, this is not information, but organizational noise.
- Information has value to the receiver.
- The receiver has a purpose for the information.

Discovering and capturing information flows is most easily accommodated by disciplines that study and model the composition and movement of data structures. These disciplines are often recognized as *process modeling, data modeling,* and *workflow analysis.* Information is generally easy to identify in your organization as it appears in tangible forms such as paper forms, transmissions, e-mails, computer displays, phone calls, and reports. A process map allows you to follow the trail of information through your enterprise.

Keep in mind that not all information is moved on physical devices; some of it travels through the spoken word. Next time you visit any business, see how much information is flowing around you. If you could actually see it, it might look like the streams of information popularized by the movie *The Matrix.*

When information is received from a sender, it is intended to entertain (think novels, music, movies, or TV shows), educate or alert, or stimulate action. In a business setting, information is generally used to trigger a decision by someone in your organization. Although there are exceptions, few organizations spend much of their information resources entertaining their employees or customers. They use information to activate a formal decision-making process.

Decision making

Decision making is the wise conduct an organization has established over time that is intended as a response when information is received. Decision making falls into three decision categories: *response, policy,* and *procedure.* Decisions may be simple or complex and they may be based on both tacit and explicit knowledge.

The response decision is simply determining if any action is needed based on the received information. Watching the daily weather forecast is an example in that we may or may not act on that information. In some cases we mentally store it away for future decisions or, if the conditions are really bad, we may seek cover or add extra layers of clothing. A great deal of business information falls into this category. Vendor product updates, phone calls alerting us that an employee is ill and will not be in for work, and proposed government or industry regulations all serve to alert us to something and give us the opportunity to respond. When we respond, we initiate either policy or procedural decisions.

Policy decisions define what the organization should do in response to a specific type of information that comes from a specific source. This information is usually initiated after a series of known events have transpired. A scenario presents itself frequently enough that your organization has defined the sequence of decisions and actions that should take place to respond to the information. Policy decisions are best documented in a way so they are independent of who will perform the work, where the work is performed, or what type of technology will be used to complete a total organizational response to the incoming information. When faced with a patient requesting a prescription refill request (information), the staff of a clinic should know what must be done (policy decision) regardless of who is actually taking the call, what physician's office the employee works in and whether the employee will do the task manually or use some form of electronic assistance. Policy decisions represent "what" must be done.

Procedural decisions are unique to a specific group of people, department, or technology implementation. They are "how" a set of policy decisions will be carried out. Procedural decisions should be consistent with an organization's policy decisions and accomplish the intent of the policy. If differing procedures are applied to a common set of policy decisions, it is possible to compare them for effectiveness and efficiency. In the prescription refill example, different clinical staff may apply their own series of steps they will use to complete the refill. But they must all satisfy the requirements of the policy decision. Because they have a common base, two clinics may be compared to see who really has the best procedure.

Explicitly capturing any form of decision making allows organizations to avoid previous mistakes and capitalize on sound business directions. It enables the creation of consistent actions when responding to internal needs or a world outside their borders.

Understanding the decision making of an organization is a daunting challenge. Gathering, capturing, and refining an organization's decision making is a complex process as most of this knowledge is held in fragmented subsets by people from far-flung organizational units. Furthermore, much organizational decision making is undefined, resulting in rampant, inconsistent, and sometimes invalid, personal interpretations. Some decision making will inevitably remain in the domain of undefinable tacit knowledge (wisdom, judgment, and intuition), however, sound knowledge discovery methods offer the opportunity to capture this elusive intellectual asset.

Conclusion

So now you know why I lead off the discussion of knowledge processes with knowledge discovery. Best practices, methodologies, and techniques used in your organization to uncover hidden data, information, and decision making are the keys to understanding and then capturing vital organizational knowledge. If the assessment of your personal and organization's discovery methods is lacking, this is the time to take action. I owe much of my career to adopting formal methods for knowledge discovery, even when I didn't realize that was what I was doing.

chapter seven

The knowledge retention policy—Level one

> Explicitly recognizing knowledge as a corporate asset is new, however, as is understanding the need to manage and invest it with the same care paid to getting value from other, more tangible assets. The need to make the most of organizational knowledge, to get as much value as possible from it, is greater now than in the past.
>
> **Thomas H. Davenport and Laurance Prusak**

In earlier chapters of this book, I made the case for finding a way to conduct knowledge asset management. I also alluded to the development of processes and templates for creating a knowledge inventory to help with the knowledge retention practice. My research with real organizations resulted in the design of a *knowledge retention policy* or KRP, which is a formal, written document that declares intellectual properties considered to be significant organizational assets.

The KRP recognizes both explicit and tacit knowledge. It describes knowledge in much the same way that a physical asset management system functions. Just like trucks, computers, and buildings are not stored in an inventory of physical assets, the KRP similarly describes organizational knowledge and then explains how to find more information about it along with the location of the actual knowledge asset.

The primary justification for a KRP is to enable an organization to know what is known. But it goes far beyond that. A KRP does the following:

1. Alerts your organization to knowledge that may be at risk due to a pending resignation, retirement, or staff reduction.
2. Is a vital resource when interviewing candidates for a new position. Instead of trying to match a candidate to a job description, you can select the specific knowledge assets that are held by others in a similar position to examine the qualifications of a new candidate.

3. Can help identify knowledge that may no longer be valid. Some collections of data, information, and decision making were based on assumptions of a previous business era and they need to be removed or changed.
4. Identifies areas where knowledge needs to grow and provides a structure for the new knowledge assets.
5. Identifies the experts for each collection of important knowledge, providing your organization with a map of who to call when help is needed.

The KRP is composed of two main components, a general management statement and a knowledge assets inventory. I have further divided the KRP into two levels of information. Level 1 declares the existence of significant knowledge assets along with informative descriptive information. Level 2 provides definition, using specific details to clarify actual content. In this chapter, I describe creating a Level 1 KRP and then explore the Level 2 in the next chapter.

Simple worksheet templates to support the KRP are provided in the appendices of this book. These working templates are intended to help you collect base data that can then be supported with a wide variety of software products, from word processors to formal repository tools. Although you have many options for actually creating this document, make sure you keep it simple and accessible. This is a living document that members of your organization will use frequently to find knowledge assets. And if KM is working in your organization, they will be updating information in the KRP on a regular basis.

General management statement

The general management statement explains the purpose for the KRP and establishes the organizational vision and expectations for knowledge management efforts within the organization. It also identifies the vocabulary and values used to score items found in the knowledge assets inventory. Specifically, I use a range of values to clarify the importance of each knowledge asset to the organization (vital, important, convenient) along with the current state of knowledge transfer. I restate these values in the footer of the knowledge assets inventory component, and I explain the intent behind the values along with the criteria of their use in the general management statement (well defined, limited definition, undocumented).

The general management statement also explains the knowledge transfer mechanisms, or KTMs, that will be used to move this knowledge. The KTMs will be further defined in this chapter. Lastly, the general management statement describes the way the knowledge assets inventory will be structured. You will find that most of the general management statement can be copied from KRP to KRP. The main objective is to explain what people will see in the knowledge assets inventory and to establish your management's commitment to this product (more on all of these elements as I explain the KRP in greater detail).

I recommend that the general management statement include the names of senior decision makers who are supporting the creation of the KRP. In Appendix B, I have supplied a template you can use. You and your leadership should craft a meaningful statement of support for the creation, use, and maintenance of the KRP and place it here.

Knowledge asset inventory

Immediately following the general management statement is the knowledge assets inventory. This KRP component is comprised of the *enterprise area* or *knowledge domain* that will be examined by this KRP, the *knowledge areas* that are recognizable to the users of the document, and the *knowledge topics* that support each knowledge area.

The most important component of the knowledge assets list is the definition of the *enterprise area* that will be addressed by the KRP. When creating this scope statement for the KRP, be sure to limit the KRP to a domain of your organizational knowledge that is feasible to accomplish. Companywide efforts become overwhelming and unfinished. The enterprise area should identify the portions of the total organization that are being studied. Equally significant is to exclude clearly any specific components that are not included.

When selecting the enterprise area for your first KRP, start with a study domain where the knowledge experts welcome the process. This will result in a higher-quality initial product. However, if you already know that a specific area of your organization is at great peril due to pending knowledge loss, you may want to create your first KRP to formally document that reality.

In a trial project, I created a KRP for the Tulsa Police Department. I had provided management training for this organization over the years and was familiar with their operation. I selected them to implement a trial KRP because I had the direct support of the senior executive, the chief of police. The Tulsa Police Department was staffed by over 800 sworn officers and was the only major city police department in the United States that required a college degree before joining the force. With a culture that emphasized education and experience, it was a great target for defining knowledge assets. The department had three deputy chiefs and a number of majors and captains with specific jurisdictions and responsibilities. The Tulsa PD KRP Enterprise Area stated:

> *This Knowledge Retention Policy identifies the Intellectual Assets considered significant to the operation of the Tulsa Police Department. It also includes activities that directly support TPD. It does not include the general operations of the City of Tulsa. This study also excludes the Information Technology support for TPD as well as the functions of the Animal Control Officer.*

This statement made it clear to all team members and reviewers what was included and what was excluded from our domain of study. By clearly stating our scope, we were able to minimize unrealistic expectations while also identifying the subject matter experts we needed to interview during the project.

You may find it useful to create a high-level taxonomy that shows all of the major knowledge domains in your total organization prior to locking down your enterprise area. For example, we needed to understand where the police department fit into the total City of Tulsa structure. Describing the total organization gave us further clarity on what we would and would not study. It is often helpful to include an illustration showing the total organization and the selected enterprise area. Participants in the process are less likely to stray outside the target conversation.

Before you go any further in the creation of a KRP, make sure you have agreement on this statement of KRP scope. It will save you a great deal of wasted time later, it will help you see where your scope fits into the larger scheme of things, and it will help you focus on the right knowledge areas and knowledge topics.

Knowledge areas

Knowledge areas are groupings of organizational knowledge that are easily recognizable to anyone familiar with the selected enterprise area under examination. They provide a natural launching point for anyone seeking a specific type of knowledge detail. The classification of knowledge areas is arbitrary and serves only to provide a first-order decomposition and organization to the more detailed knowledge topics. Knowledge areas may be formal organizational units or important functions performed by a subset of the organization.

In the Tulsa Police Department example, knowledge areas included the Detective Division, Forensics, Headquarters, Homeland Security, Training, Uniformed Divisions, and many others. Due to the complexity of operations within Headquarters, that organizational unit was further decomposed into distinct knowledge areas of Personnel, Booking, Administration, Property Room, and Records.

The key to identifying knowledge areas is to engage the senior decision makers for your enterprise area. They should be so familiar with the domain of study that they can provide this structure with little effort. Be cautious, however, of just following the current organizational chart. It is not unusual to discover significant redundancy across similar organizations. This can be resolved later. For example, the initial KRP for Tulsa PD listed three Uniformed Divisions. That is the way they were politically and geographically structured. As we dug deeper, however, we discovered that all three divisions did the same things with very little variation. As a result, we combined them into one "Uniformed Division" and listed subject matter experts from each location.

Not all knowledge areas can be found on the organizational chart. Special programs, designed to support the entire organization, should be listed as well. For example, I helped the Tulsa Police Department create an ideation program titled "Managing Law Enforcement Initiatives (MLEI)." This program was used to generate new ideas to solve problems or provide new capabilities. If a new idea was approved, it then entered the domain of formal project management. As the MLEI processes cut across the total organization, it was recognized in the KRP as an independent knowledge area.

You should also be careful not to become so attached to a current organizational structure that you limit needed restructuring of the enterprise. The KRP should be a tool to think through reorganizational efforts, not hinder it. The identification of knowledge areas is arbitrary, but don't be surprised if you see it change during the process. Initial definitions are often based on traditional established structures found in the organization. Once knowledge topics are added to the structure, it is common to see a new organization of knowledge areas emerge. Once you have identified a knowledge area, it is helpful to provide a short narrative description along with the name or names of the people who are organizationally responsible for this portion of the enterprise. Keep this information updated as the organization changes.

Knowledge topics

The actual definition of a knowledge area occurs when associated with a specific set of knowledge topics. *Knowledge topics* represent known collections of work processes, methods, techniques, standards, responsibilities, and information that are significant to the organization. Knowledge topics may be organizational programs, business processes, business data, and even entire application systems.

In the Tulsa Police example, the Detective Division knowledge area listed knowledge topics of Auto Theft, Burglary, Child Crisis, Cybercrimes, Diversified Crime, Exploitation, Family Violence,

Financial Crimes, Fugitive Warrants, Homicide, Major Crimes, Robbery, Sex Crimes, Sexual Offender Registration, and Sexual Assault Nurses Examiner. Quite a list, isn't it? It makes you wonder about the law enforcement organizations where you live. How do they compare to this list? I wonder how many of these could the city mayor describe?

Keep in mind that this was just for one of the 25 knowledge areas identified for Tulsa PD. This list didn't take months to compile, just a few days. This is what the people in the Detective Division do and think about. This is how they train their officers. They have formal collections of organizational knowledge (data, information, and decision making) to support every one of these knowledge topics. In some cases, the knowledge topics are supported by legal policy. In others, they have created what they call TOGs or tactical operational guidelines. TOGs are intended to remind officers of how to perform a skill they may not use with regularity. It is explicit knowledge created to foster wise conduct. For example, when the dog team is called in, the first action of the officers on the scene is to "get in your patrol car." The dogs are trained to run down bad guys and they don't always recognize the uniform. A badge doesn't impress them very much.

When we were creating the lists of knowledge topics for each of the knowledge areas, some of the participants were not sure where to start. I suggested that they simply use this forum to explain to others what it is that they do. Do your customers really know what you do and what you know? How about your own management? Do they understand what goes into the major knowledge areas of your environment? Here is a way to go on record with what it actually takes to run a knowledge area within an enterprise area.

Obviously, some of the knowledge topics will have more universal meanings than others. You and I could take a pretty good shot at Burglary (no pun intended), but what is included in Child Crisis or Diversified Crimes? That is where the remainder of the template for knowledge topics comes into play.

Once identified, a knowledge topic must be defined with 1) a description, 2) the identity of the knowledge expert or experts within the organization, 3) the importance of this knowledge topic to the organization, 4) current transfer status, and 5) the knowledge transfer mechanisms that are needed to share this knowledge in the organization.

The description is simply a short paragraph or two that will help future readers understand the general intent and reach of the knowledge topic. For example, the description of Child Crisis was

> *Responsible for the investigation of sexual abuse of children under the age of 11, physical abuse of children under the age of 18, infant fatalities, abandoned and endangered*

children, children found in methamphetamine labs, and children in need of supervision.

Tulsa PD made a distinction between Child Crisis and Exploitation. Exploitation was described as the following.

> Juvenile exploitation includes sex crimes against children ages 11–15, child stealing, Amber alerts, lewd or indecent proposals to a child under 16 years of age, child pornography (noncomputer), and runaways.
>
> Senior services include offenses against a person over the age of 62 when age is a factor in their victimization. Investigations include adult in need of supervision, elder financial crimes, forgeries, embezzlement, kidnapping, sex offenses, strong-armed robbery and burglary.

Diversified Crimes was explained as

> Responsible for the investigation of assault and battery, larceny, malicious mischief, hit and run, traffic fatalities, and other miscellaneous crimes.

For the record, I did not create any of the above descriptions. They were copied directly from existing Tulsa PD manuals.

Knowledge expert

Associated with each knowledge topic is the name or names of the knowledge experts for this content along with their contact information. These are the people who are responsible for the full definition of the knowledge topics and making sure the specifics are maintained. One of the responsibilities of these experts is to judge the importance and transfer status of the knowledge topic.

Organizational importance

I rank organizational importance of a knowledge topic as "vital," "important," or "convenient." A "vital" knowledge topic represents knowledge that if lost will cause devastating implications to the organization including lost capabilities or complete failure. A "vital" designation may also represent knowledge at risk with a short window for capture. This type of knowledge topic is often the result of tacit knowledge held by a limited number of current employees. Designating a knowledge topic as "vital" raises attention to the significance of capturing, retaining, and transferring this knowledge.

"Convenient" knowledge topics represent processes or information useful to the organization but that may be recovered with limited effort. Although it is important to retain and transfer "convenient" knowledge topics, they represent a lower risk of loss to the organization. Between "vital" and "convenient" will be "important" knowledge topics. The loss of this knowledge will compromise the organization and re-creating the lost knowledge will be difficult and expensive, but not devastating.

It will come as to no surprise to you that Child Crisis, Diversified Crimes, and Exploitation were all rated as "vital." As a citizen and parent, I would agree.

Transfer status

Rating the importance of a knowledge topic must then consider another factor. What is the current transfer status of this set of organizational knowledge? In other words, how well prepared is the organization to make sure this knowledge is passed in its entirety to new employees responsible for this knowledge topic?

I gave the knowledge experts three choices for this attribute. They could rate each knowledge topic's transfer status as "well defined," "limited definition," or "undefined." "Well defined" indicates that formal processes are currently in place to actively transfer the knowledge from one individual or organization to another. "Limited definition" indicates some of the pieces are in place, and "undefined" indicates there is little, if any, formal process for moving this knowledge.

Obviously, the purpose of this rating is to consider it in relationship to the importance of the knowledge topic. "Vital" knowledge that is currently "undefined" requires immediate organizational attention. "Limited definition" for an "important" knowledge topic that is held by a retiring employee should also result in a strategy to prevent knowledge loss.

The definition of each knowledge topic includes one final piece of information. The description of a knowledge topic is completed by identifying how the described knowledge should be transferred.

Knowledge transfer mechanisms

When describing the knowledge transfer process in Chapter 4, I suggested there are at least six proven ways to move knowledge. I refer to these strategies as knowledge transfer mechanisms, or KTMs. Some are positioned toward explicit knowledge whereas others are more effective with tacit knowledge. When considering a specific knowledge topic, you will usually identify more than one KTM to transfer the important knowledge effectively. Recognized transfer mechanisms include:

- *Documentation:* One means to capture and transfer organizational knowledge is using physical or electronic artifacts. These artifacts include all datatypes such as text, graphics, images, audio, and video, on any medium. Emerging artifact types will include the electronic composition of a physical item that may be reproduced using holographic or reconstruction technologies. Replicator technology is no longer the sole domain of science fiction. The use of documentation to transfer knowledge indicates the knowledge is explicit. In the Tulsa PD example, their policies and TOGs would be included in this knowledge transfer mechanism.
- *Training:* Organizational knowledge is often transferred using an educational process. Training may be comprised of a formal educational curriculum or specific task training. For the Tulsa PD, they send new recruits to a six-month academy supplemented by in-service training programs for specific skills.
- *Apprenticeship:* Complex significant knowledge is often transferred using the relationship between an expert and apprentice practitioner. Apprenticeship implies a dedicated, sustained transfer process. This practice is a proven way to convey complex tacit knowledge to an apprentice. It has a rich history and is still used by many crafts to teach new practitioners the finer points of a skill. At the completion of the Tulsa Police Academy, each officer is assigned to a Field Training Officer who rides with the rookie for several months. It is here that a new officer learns the important lessons that are difficult to teach in classrooms or simulations. A similar process is used to train new physicians during residency.
- *Mentoring/Coaching:* Often as a supplement to other forms of knowledge transfer, mentoring provides an ongoing benefit. Mentoring or coaching includes identifying people who are available to provide advice and assistance to someone performing a new task. When opening a new store location, a major convenience store operation assigns a mentor for each specific location. That mentor works on-site with the new store manager during the initial opening. He or she remains available, 24 hours a day, for several months. Portions of the mentor's compensation are based on how well the store performs after the mentor is no longer on site.
- *Cross-Training:* Many organizations enable knowledge transfer by placing less experienced people with task experts in a "job shadowing" process. This is a valuable strategy to create redundant coverage for specific skills and responsibilities. When the chief of police in Tulsa is away for any extended period of time, command is passed

over to one of the deputy chiefs or majors. This act ensures that capable command is always available in the department.
- *Communications:* A great deal of organizational knowledge is transferred using formal and informal communications. Formal communications include professional societies, committees, conferences, job-related websites, and reference books. Informal communications include social networks, social events, and chat rooms that will support the transfer of specific knowledge.

KRP activities

Capturing this information about each of the knowledge topics is not particularly difficult. Nor is it all that time consuming. Although every KRP I've created takes a slightly different path, here is a general plan for pulling the Level 1 document together:

1. *Define the Enterprise Area.* People usually already have some idea of why they want to engage in building a knowledge inventory but they don't know what is involved. They may also be trying to convince their management and peers that this is a good idea. I find it useful to conduct a joint session where all interested parties, including management, can first learn more about the KRP and then assist with defining a target enterprise area. I use a standard presentation to give an overview about KM and the role of the KRP. To supplement that presentation, I provide them with the templates we will use and a sample of a previous KRP. I then lead them through defining the scope of their enterprise area by establishing what is "in" and what is "out" of scope. The deliverable from this working session should be a concise statement of the proposed enterprise area. Don't hold on to it too closely, as it will likely change during the process. At the close of this session, I walk the group through a handout showing the general sequence of upcoming events and request that we schedule times that will accommodate creating the KRP Level 1. This will take some planning on their part. The balance of this list will help. As a final note, I like to schedule this opening session for two hours to half a day. Consider running this in the morning to avoid the postlunch lull.
2. *Define the KM Vision Statement.* There is additional information on this deliverable elsewhere in this book and the template is in Appendix A, but I think it is important to include it in this specific workflow. The KM vision statement enables your organization to clearly establish its expectations for knowledge management and to recognize efforts that are currently underway that will contribute to

this effort. Starting with the enterprise area established during the previous session, plus any subsequent refinements, I walk the participants through the notion of the KM beliefs and processes (Chapter 3 and 4). While projecting the template on the screen, I ask the group to personalize their expectations for each of the KM processes. I find that starting with formal methods and strategies is more tangible than asking them to take ownership for the KM beliefs. That will come later.

For each KM process, I also ask them to identify activities that are currently present in their organizations and those that are planned. With the KM process in a column, it is easy to establish a row for each participating organizational group. For example, you may find a technology group that has created a procedure they use every time an internal client wants to add new software to the system. Or your marketing department may have plans to build a new client on-boarding tool that will provide needed information for affected groups in your organization. Both of these would fall under knowledge discovery. I use one color to indicate something that is already in place and another color for a future contribution. The moment you have one or two tangible examples on the screen, expect others to catch on and begin to interact.

A major benefit of this exercise is that the total group will become aware of KM activities around them. Knowledge capture and retention are easy. That is why everyone is in the room. But you may find other efforts that tie into what you are doing.

Once you have done an assessment of current and future KM processes, move on to the KM beliefs. Simply ask people, based on what was identified about the KM processes, how to restate the belief statements provided in Chapter 3. Even if there are no existing statements such as these in your organization, this is your opportunity to set the standard. Once completed, the KM vision statement, along with a work plan for creating the KRP, can be presented to your management for approval and authorization to move forward.

3. *Propose Knowledge Areas/Assign Knowledge Topics.* For the next few sessions, your efforts will alternate between refining knowledge areas and associating knowledge topics. Begin this process by collecting a list of proposed knowledge areas. Keep in mind that there is no formula for this process. You are looking for the big categories of the work done within the defined enterprise area. Provide a couple of suggestions and let the team take it from there. Consider using sticky notes that can be moved around. Add clarity to a knowledge area by attributing to it a list of knowledge topics. If a knowledge area is too granular, it will be difficult to assign knowledge topics. If the knowledge area is too broad, you will find a large number of

knowledge topics. Look for distinct subcategories of these knowledge topics and use this information to refine your knowledge areas. Continue this process until the participants are pleased with the total structure. Make sure you complete this activity by fully describing each knowledge area and identify the person or persons who are organizationally responsible for it, the knowledge area owner. I also ask for recommended subject matter experts (SMEs) for each of the knowledge topics.

4. *Describe Knowledge Topics.* Once the general structure is reasonably stable, it is time to divide and conquer. I find it most effective to create the specifics that explain a knowledge topic with the designated knowledge area owner and the SMEs identified for the knowledge topic. It is also easier to schedule times with this smaller group. Start by creating a meaningful description of the knowledge topic. Add the organizational significance and transfer status for each knowledge topic. Update the list of SMEs and then identify the knowledge transfer mechanisms that are or will be used to move this knowledge.

5. *Review Completed KRP Level 1.* Consolidate the various knowledge topic worksheets into a single KRP Level 1 document. Distribute the result to all of your participants for review. I like to work with one of the more progressive teams to build a handful of sample Level 2 documents. They will help the organization understand the value that may be derived from this information along with a projection of the time required to complete the Level 2 materials. Schedule a review session and walk everyone through the result. Ask for proposed revisions prior to submitting the KRP for formal approval.

6. *Review with Senior Management.* Once you have the complete Level 1 version of the KRP (it is never actually done), it is time to review the results with your senior management. Avoid trying to explain how it was created. Instead, focus on what the document can tell them. Point out areas where the organization is fully prepared for knowledge transfer and where there are weaknesses. Offer suggestions on how knowledge gaps may be filled. Also use this time to display sample Level 2 documents and recommend specific "at risk" areas where that effort is most significant. Be prepared to provide timelines and resource estimates.

7. *Determine Implementation.* A final task before moving on to the Level 2 elements of the KRP is to determine (1) what technology will be used to store the KRP, (2) where the KRP will be retained, and (3) who will be responsible to keep it current. The initial worksheets I use to collect KRP data are based on word processing templates. Although some organizations choose to keep the document in this format, you should also discuss the potential for storing this material

in a formal repository product. The selection of technology for storing the KRP may provide guidance on where it will be kept. If the KRP is being captured on a shared drive, you will want to make sure all of the participants have access to the information. Last, make sure you have identified a custodian for the KRP. You will need to construct a simple process for making revisions and a person who will facilitate these updates.

Appendix C provides you with a general worksheet that may be used to capture most of the information discussed in this chapter. Feel free to revise it to fit your needs.

Conclusion

What I have described here is a very effective knowledge inventory. It is also the Level 1 version of the KRP. The Level 2 study involves creating the specific details, or metadata, needed to validate each of the knowledge transfer mechanisms. You will find those details in the next chapter.

But don't overlook what you can accomplish with just the Level 1 document. A Level 1 KRP allows you to clearly identify the organizational knowledge required to perform the work within a specific enterprise area. It describes both explicit and tacit knowledge along with who currently supports this knowledge as the subject matter expert. The Level 1 examination also allows your management to recognize and respond to real risks of legitimate knowledge loss.

If you are still undecided about investing the time to create a Level 1 KRP, the most significant application of this work is something I've not even mentioned to this point: using the KRP's structure to guide the creation of a knowledge portal. If you have any expectations of implementing KM as a normal way of doing business, you will want to create a common technology system where your people may store significant organizational knowledge. The definition of enterprise areas and the top structure of the KRP gives you the structure for a knowledge portal. Much of the descriptive information may also be added to help explain the intent of the Level 1 KRP elements.

So, start here. Select a reasonable enterprise area. Build a team to help you identify the knowledge areas and knowledge topics. Fill in the specifics. Before you know it, you will have a true inventory of your knowledge assets.

chapter eight

The knowledge retention policy—Level two

> It's the little details that are vital.
> Little things make big things happen.
>
> **John Wooden**

Once you create a Level 1 KRP for a specific enterprise area, it is time to move forward to the Level 2 study. But before we go on, take a minute to reflect on the significant value a Level 1 study gives you. Not only will it help you formally recognize the explicit and tacit knowledge assets in a segment of your organization, the KRP also delivers a tangible result in a reasonably short period of time. In addition, the people who participated in the process will have a much clearer understanding of knowledge management and how to conduct similar activities for other enterprise areas. If you use the same approach to building other KRPs, you will be able to combine them into a master version over time. The Level 1 document declares the knowledge assets of your organization, the Level 2 material defines it.

The Level 2 study offers significant value as it helps you uncover and document the details behind each of the knowledge transfer mechanisms (KTMs) listed for each knowledge topic. This is not difficult, but the process may demand substantial time to accomplish. In fact, the Level 2 study will require such constant revision that it is never really done. The dynamic nature of the Level 2 KRP content is not due to the template but the nature of your organizational knowledge.

As with the L1 content, a Level 2 KRP should be completed using a set of standard worksheets. I have provided a sample master template for the Level 2 study in Appendix D. You may find it useful to put a tab there so you can jump back and forth between this chapter and the template, where there are a series of questions that should be answered to validate the appropriateness and accuracy of each knowledge transfer mechanism listed for a knowledge topic. In most cases, multiple KTMs are recommended to transfer the knowledge of a specific topic. The template supports all six KTMs described in Chapter 7. Remove unneeded segments for a specific knowledge topic.

In this chapter, I suggest a set of attributes (metadata) for each knowledge transfer mechanism. This material should be used as a starter list, not the final result. You will want to engage a community of practice comprised of people from various professions to help you refine the information you wish to keep regarding each KTM. This may also be the time to invest in a repository product specifically designed to capture this type of content. The purpose of these metadata is to enable anyone who might become responsible for a knowledge topic to utilize proven materials and not waste time and money rediscovering the knowledge.

Documentation

Clearly, physical documentation of any type is the most tangible of the knowledge transfer mechanisms. The specific attributes I recommend for this KTM are the following:

- *Document/Record Name:* Document the group name for a distinct set of records that transfer this knowledge.
- *Record Type:* Identify the type of record for this group. Type includes text, graphics, video, and audio.
- *Location:* Specify where this group of records will be physically located. This includes URLs, software products, or filing location.
- *File Naming Standards:* Describe the composition for the name of this document that will make it easy to store and find in an electronic repository. When possible, engage a community of practice with primary responsibility for this specific document when creating the naming standard.
- *Technology:* Specify the technology that will be used to capture this record type.
- *Update Type:* Identify whether this record is static or dynamic. Static records are locked once created whereas dynamic records are updated on a regular basis.
- *Revision Schedule:* Specify how frequently this set of records should be reviewed and updated.
- *Revision Responsibility:* Name the person who is responsible for reviewing this set of records.
- *Access Rights:* Determine how this set of records may be accessed. Who is allowed to see the records? Will they be for internal use only or will they be provided to an external audience? What registration, if any, is required to gain access?
- *Security:* Specify the type of security needed to protect this record.

- *Retention Term:* Identify the length of time this set of records should be retained.
- *Disposal:* Specify how this record should be archived or destroyed.

There are many potential documents that must be considered for this type of knowledge transfer. You will need to create a triage process to determine which of the candidate records provide value to the organization and which just add clutter. Examine each candidate record type to establish how it supports the defined knowledge topic in your KRP. Be aware that some records create an organizational liability when retained. It is a good idea to have your compliance and legal representatives review the list of records you plan to retain.

You will also find there are many additional metadata candidates that could be useful for describing each named document. I often list all of the documents in the template provided and then create a separate worksheet for each document to explore the detailed metadata. This is where your knowledge management and enterprise content management (ECM) strategies intersect. ECM products are intended to implement document retention policies and use a wide range of technologies to capture electronic versions of the records. Many vendors stand ready to help your organization design solutions for document and records management.

A major distinction your KM focus brings to enterprise content management comes from the knowledge retention policy, specifically the definition of a specific enterprise area. Most ECM efforts are launched with the intent to identify document and record retention needs for the total organization. Defining detailed metadata for such a broad scope overwhelms even the most ambitious teams. Before long, the ECM project simply becomes an effort to store physical records in a repository. When performed for a specific knowledge domain, ECM activities take on a much narrower and more achievable scope. The Level 1 information also puts the more detailed elements into context.

Training

Training is a proven effective way to transfer both explicit and tacit knowledge. Some training programs provide specifics on individual tasks whereas others explain a general concept and philosophy and leave application to the student. When training of any type is proven effective preparing for a specific knowledge topic, it is important to record the following:

- *Training Name:* Provide the name of a formal training program (individual course or series) that facilitates this knowledge transfer.

- *Training Description:* Summarize the learning objectives, delivery means, time requirements, and any other information you believe will communicate the contents of the training.
- *Training Duration:* Identify the amount of time required to complete the training program.
- *Training Costs:* Note the costs that will be incurred to obtain this education and any related certifications.
- *Trainer or Vendor Information:* Include contact information that can be used to contact the actual trainer or training vendor. This will enable locating course descriptions and registration information.
- *Training Prerequisites:* Specify the previous education or experience level that should be completed prior to this training.
- *Resulting Certifications:* List the degrees or certifications that will verify successful knowledge transfer.
- *Performance Measures:* Describe the performance measures that should be met to validate this knowledge transfer?

Apprenticeships

For centuries, apprenticeships have proven effective for transferring challenging knowledge, especially tacit understandings that resist formal expression. When the knowledge topic represents intellectual assets that are vital to an organization, the increased time required for an apprenticeship may remain the best way to communicate the knowledge. What differentiates an apprenticeship from mentoring or coaching is the amount of time involved and intensity of the assistance. An apprentice is under constant supervision for an extended period of time, sometimes years. A mentee only requires the advice of a mentor on a limited basis. When proposing an apprenticeship, include metadata for the following:

- *Apprentice Qualifications:* What skills or education is needed by an apprentice?
- *Apprentice Selection:* What are the suggested selection criteria for an apprentice?
- *Selection Process:* What is the suggested selection process for an apprentice?
- *Apprenticeship Term:* What is the recommended length of apprenticeship?
- *Apprenticeship Completion:* What certifications or performance capabilities indicate the successful completion of an apprenticeship?

Mentoring/coaching

Many knowledge topics can benefit from the support of short- and long-term mentoring. Often used in combination with other knowledge transfer mechanisms, this transfer process provides help from an expert when needed. Useful descriptive information includes the following:

- *Mentor Qualifications:* What are the criteria for a mentor?
- *Mentor Commitment:* How much total time and incremental time will be required of a mentor? What is the total length of a mentor relationship?
- *Mentor Recognition:* What types of recognition or reward are available for the mentor?

Cross-training

Cross-training is recognized as a valuable way to provide back-up practitioners for any set of knowledge topics. Similar to the metadata for apprenticeships, you may want to capture the following:

- *Trainer Qualifications:* What are the criteria for the trainer?
- *Trainee Qualifications:* What are the criteria for the trainee?
- *Training Frequency:* How frequently should the cross-training sessions occur?
- *Training Term:* How long should each cross-training session last?
- *Training Completion:* What certifications or performance capabilities indicate successful completion of a cross-training effort?

Communications

With the wealth of information available from so many sources, it is very helpful to have a knowledge expert recommend a collection of communication sources that will make performing the work of a knowledge topic more effective. Communications may recommend professional societies, committees, conferences, periodicals, blogs, websites, and books. It could even include staying in contact with specific people who provide valuable insights into the knowledge topic. Use these metadata to alert future practitioners to the following:

- *Communication Name:* The type of communication process recommended.
- *Communication Type:* Professional organization, committee, publication, website, social network, and so on.
- *Information Source:* URL or address to obtain additional information.

- *Communication Contact:* Name of person to contact about this communication.
- Communication Frequency: How often this communication occurs or how often it should be reviewed.

Conclusion

The foundation for a knowledge retention policy is found in the Level 1 study, however, the details provided in the Level 2 documentation provide the true substance. When creating a Level 1 KRP, it is a good idea to create a Level 2 document for a few significant knowledge topics. They serve as a sample to gain buy-in from your management to approve the required time to complete Level 2. It will also provide an example for others to use when creating other Level 2 documents. I also recommend that you ask the subject matter experts listed in the Level 1 study to compile this Level 2 information. They should know this information well or know how to get it. It is equally important that these experts commit to reviewing and updating this information on a regular basis. The Level 2 information provides the most secure means to ensure the transfer of vital knowledge assets.

Lastly, although I have provided you with suggested templates to capture these metadata about the content of your selected knowledge domain, please take the initiative to make changes to these basic forms. They are provided simply as content guides.

chapter nine

A model for managing organizational knowledge

> Some people prefer to learn by manipulating models and others by reading.
>
> **Bill Gates**

I hope this book has given you an increased appreciation for a broader view of KM (based on the KMBPs in Chapters 3 and 4), a focus on the elements of organizational knowledge (in Chapters 5 and 6), and a formal process that enables the creation of a knowledge inventory for specific domains of your organization (KRP in Chapters 7 and 8). You may also have a fresh understanding of the challenge ahead of your organization to put all this into place. If this material makes sense to you, there is still one major missing component: a general model to guide the implementation of a sustainable strategy for knowledge management. Enter the KIPPAR model.

KIPPAR Model

The KIPPAR model I propose integrates the elements that I believe are needed to keep your KM efforts growing and current. Specifically included are the following:

- A *knowledge inventory* (*KI*), or knowledge retention policy, that (1) establishes the organizational intent for meaningful knowledge transfer and (2) provides an asset inventory of significant knowledge elements. This inventory recognizes both tacit and explicit knowledge.
- Three pillars of (1) *projects (P)* that establish "when" organizational efforts are most likely to use existing knowledge and create new knowledge, (2) knowledge producing *processes (P)* that explain "how" predictable artifacts should be created, and (3) valuable *artifacts (A)* that identify "what" knowledge artifacts exist within the organization offering long-term value. All three of these pillars have independent characteristics but work in an integrated man-

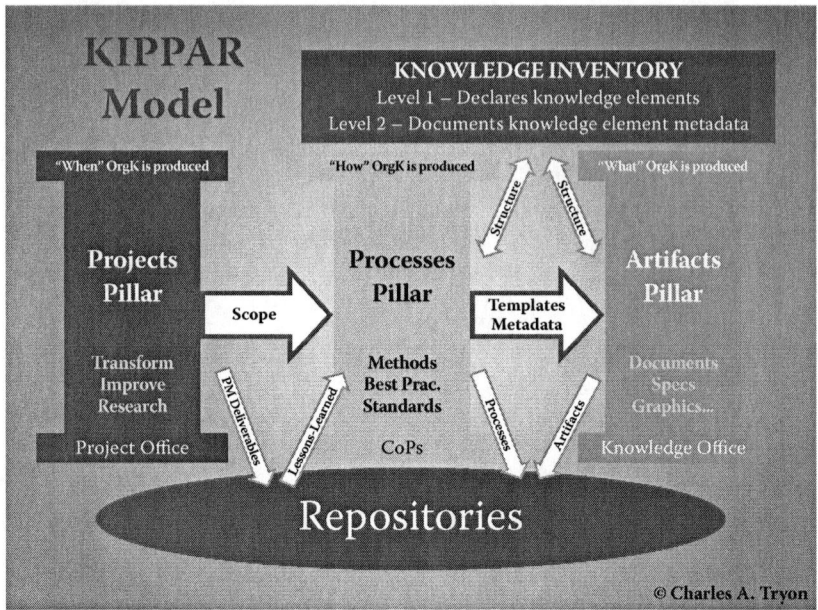

Figure 9.1 KIPPAR model.

ner to produce new knowledge or make refinements to existing knowledge.
- A deliberate collection of *repositories* (R) that will capture the resultant knowledge of an organization.

The purpose of this model is to provide you with a general strategy that guides how you can implement a sustainable approach to knowledge management. The following pages examine each of these components and provide recommendations for a practical implementation. The elements of this model (see Figure 9.1) are each reviewed below.

The knowledge inventory

When viewed through the KIPPAR model, the knowledge retention policy provides the top-level structure to organize both tacit and explicit knowledge. This taxonomy is based on how the business views itself instead of some arbitrary classification scheme. Chapter 7 covers this inventory of organizational knowledge in great detail. As is shown, this structure helps you recognize and organize knowledge assets that might be missed or trivialized. Failure to create this product will leave the organization of your knowledge to trial, error, and rework.

The artifacts pillar

The most recognized pillar of this model is the by-product of the projects and processes pillars. The projects pillar indicates when knowledge will be used or created, the processes pillar captures how an organization wishes the work to be performed, and the artifacts pillar identifies what explicit knowledge is most highly valued by the organization. Among common artifacts are written documents, engineering schematics, requirements, specifications, drawings, graphical models, reports, analysis, legal findings, financial reports, research papers, designs, policies, and operational procedures. Literally, anything your organization deems significant and reusable will show up in the artifacts pillar. To determine the artifacts most valuable to an organization, qualified communities of practice should examine the natural by-products of their best practices or processes. Each recognized formal process should provide multiple artifact candidates. Look for artifacts where the results from one document become input to a subsequent document.

While examining potential assets produced by best practices, it is also useful to examine how distinct groups of users perform their functions. This bottom-up examination provides the opportunity to focus on artifacts that show the greatest real organizational value as these records have a proven use. This approach examines the information use of distinct groups of users within an enterprise. The identification of these user groups enables the examination of their information seeking and use behaviors. The results of these studies can provide valuable information on how the records should be organized, stored, and accessed later. All of these operational processes should be linked to a knowledge area and knowledge topic in your knowledge retention policy. Specifics that define these knowledge assets may be recorded in the Level 2 KRP document you create for a specific knowledge topic.

Once a class of artifacts is identified, a community of practice that understands the artifact should determine if it represents a long-term asset that will be maintained by the organization or a short-term asset that provides only temporary value. Although both may be retained, greater emphasis and rigor should be required of long-term assets.

Each type of long-term artifact should be clearly defined in a Level 2 KRP worksheet.

Due to the diversity of artifact types, their special characteristics, and the diversity of users, a variety of repository products may be needed. Technical artifacts with special formats and complex metadata will require repository products that are specific to those components and others may be retained in a common document management product.

With the number of significant artifacts that carry implications beyond their place of origin, many organizations are finding it necessary to create a governance structure to monitor the proper retention of these artifacts. A knowledge office could prove useful to sustain a knowledge management environment. It would also be charged with validating the quality, content, and maintenance of all retained artifacts.

The processes pillar

When creating valuable artifacts, your staff will utilize a variety of formal processes or best practices to produce valuable knowledge assets. These processes include grand development methodologies, technical methods, formal techniques, organizational procedures, industry standards, and recognized best practices. The processes pillar identifies how work may or should be performed as some processes are suggested and others are mandated. To enhance management's recognition of this form of organizational knowledge, significant collections of standard processes should be listed in the KRP as knowledge areas or knowledge topics. The value of formally recognizing these processes is to enable (1) greater consistency across the organization, (2) the production of more consistent work products, and (3) the collection of more meaningful lessons that may be learned from use by diverse teams.

For example, many discovery methods used on projects begin with an examination of current business practices prior to creating new solutions. Formally recognizing the steps required to create the current state views of business activities establishes a standard process that can provide consistency if used across the organization.

Defining the scope of each project (in the projects pillar) allows for the selection of specific best practices, methodologies, and standards captured in the processes pillar. Instead of allowing each project team to find their own way through an assignment, they should be encouraged to utilize proven strategies that have been defined and captured in the processes pillar. These project teams will be able to consider the strengths and weaknesses of each approach along with any special modifications their project will require. The use of repeatable processes will also help identify candidate artifacts that should be retained or updated based on their efforts.

Administration for the processes pillar should certify the effectiveness of each standard process, facilitate education in proper use, and continually refine the process based on new findings, current projects, or changing needs. Organizations should establish a community of practice for each distinct process group to monitor the use of the process and provide a vehicle for continual refinement. These champions of specific processes should also define repository needs.

The majority of these processes will be captured in documents, images, and graphics. It is vital that these processes be easy to find and retrieve.

If the processes are difficult to locate, practitioners may ignore them or create redundant versions. Some formal methodologies may be acquired in vendor-supplied repositories where other elements will require some form of document management.

The projects pillar

The KIPPAR model allows organizations to leverage naturally occurring projects by recognizing and harvesting valuable knowledge assets that are routinely used or created by these projects. Failure to recognize the knowledge value present in these projects often results in these assets being used and then stored carelessly or completely ignored.

As stated earlier, projects are organizational activities with a specific scope, assigned resources, and determined schedules that are intended to create new products and services or make revisions to existing products and services. Organizational projects are recognized in three broad categories: transformational, improvement, or research endeavors. *Transformational* projects evaluate current business activities and related information with the intent to make fundamental changes in order to achieve radical improvement for an organization. *Improvement* projects utilize the industrial engineering concepts of standardized work activities, specialized workers, and synchronized integrated work to make improvements to a business operation or information currently used by an organization. The third type of organizational projects emphasizes *research* that delivers new understandings or breakthrough concepts. Often referred to as "research and development," this project type is becoming increasingly significant to organizations as they struggle to remain viable in markets driven by impatient customers with limited loyalty.

No matter the type of project, it is clear that they all create or refine volumes of organizational knowledge that should be preserved for future operations or projects. The KIPPAR model is specifically designed to capture knowledge from projects of all types. As the KIPPAR model is predicated on harvesting organizational knowledge from projects, it is necessary to compile a list of projects being performed within an organizational area. Creating a KRP, establishing a knowledge management initiative, and selecting usable repository products all represent projects. Organizations should maintain an updated list of all projects currently underway along with projects that are pending approval. The projects pillar represents when organizational knowledge will be identified, used, and created.

The projects pillar is formalized by identifying specific information that must be created and maintained during the life of the project. Prior to launching each project, project managers create documents that define the nature of their project along with plans to perform the work. They then track issues, changes, and status during the life of the endeavor. This information should be retained in a project repository, providing access to all team members and stakeholders.

Management of the project pillar is best facilitated by a project office that provides assistance to new teams and monitors the project repository. The project office also establishes the baseline processes needed to plan and control any project. They also serve as the community of practice that recommends standard templates for use by project managers. Projects are a fluid entity for an organization as some are completed, others are cancelled, and new projects are initiated. Because of this constantly changing nature, projects are not listed individually in the KRP. The project pillar should be supported by a dedicated project portal.

The greatest contribution of the projects pillar to the KIPPAR model is the definition of scope for each specific project. *Project scope* establishes the context for a project as it declares a domain of study, domain of effort, and a list of deliverables that will be included and excluded from the focus of the project. *Domain of study* identifies the specific locations, functions, business processes, or data that the project will and will not examine. *Domain of effort* identifies the types of work that a project team will and will not perform. *Deliverables* explain the primary results or outcomes that will and will not be produced from this specific project. This information is validated with the appropriate project owner to ensure the project will address expected needs.

In the KIPPAR model, this statement of individual project scope will provide the basis for selecting proven best practices found in the processes pillar. Scope will also provide an ongoing mechanism to identify and certify new emerging processes for an organization. A second significant contribution of the projects pillar happens at the end of each project when a formal review of project results is conducted. These "lessons learned" or "postmortem" sessions examine the effective use of selected processes and the resulting artifacts for refinement within the organization. It is vital that this postproject review information be distributed across the organization, specifically targeting other current project teams and the communities of practice responsible for the processes used and their resulting artifacts.

Repository products

As described for the projects, processes, and artifacts pillars, organizations will require an integrated repository strategy that accommodates the needs presented by the organizational knowledge assets that have been identified. Prior to creating or purchasing these repositories, it is ideal to identify the creators and users for each set of information. Understanding how user groups plan to use repository products allows for better product

Chapter nine: A model for managing organizational knowledge 73

design and implementation. Most repositories define a primary structure based on the most common use of the artifact. Secondary searches are accomplished using keywords. Emerging search strategies examine actual artifact content for potential matches.

KM or ECM

Many organizations launch a KM implementation by purchasing one or more repository products and then importing volumes of content. Instead of a KM effort, this is actually an enterprise content management (ECM) effort.

ECM is a useful organizational effort. But if that is how you plan to lead the move to KM, you will encounter many false expectations and shortcomings. It is not a matter of right or wrong, but it is a matter of what you are trying to accomplish. Also known as document and records management (DRM), ECM is designed to help an organization capture explicit knowledge. This is done by providing storage for physical records, mostly electronic images or scanned paper forms. Emphasis is placed on storing important agreements and transactions. Consideration is given to organizing these records and tagging them with sufficient metadata so the records can be located should they be needed. The motivation for ECM is to protect against legal or financial challenge and to prove regulatory compliance with some added value for customer service.

KM, on the other hand, helps an organization understand the underlying reason for creating and storing records. It explains how to go about applying proven organizational processes. KM examines the behaviors of an organization, constantly seeking better ways of doing things. KM helps make meaningful judgments about what content should and should not be stored. As a result, KM must examine both explicit and tacit knowledge with special emphasis on knowledge that cannot be stored or transferred. In short, ECM is about document retention, whereas KM focuses on knowledge transfer.

When creating a knowledge retention policy, ECM may provide the technology for retaining documentation that supports each knowledge topic. The organization of the records, however, is based directly on the knowledge areas and knowledge topics responsible for their creation and use.

The most significant challenge for ECM is the size of the effort and the granularity of the target. The very thought of organizing detailed information about a total "enterprise" should trigger all types of warning flags for corporate executives. It is an overwhelming task that will likely never see the promised benefit. Global corporate initiatives often struggle with the effort required to implement anything companywide. The prudent response is to implement ECM as a by-product of your knowledge management effort, not the other way around.

The KIPPAR model recognizes the need to retain vital explicit records in electronic repositories, but the model encourages finding the reason behind the documents first. You can do that by recognizing the value of repositories, but put that topic aside for the time being. Instead, focus on the three pillars that will provide a reasonable way to uncover meaningful records to retain.

Conclusion

Although no single model can serve the needs of all organizational efforts needed to manage your knowledge, it is clear that a lack of any guidance on the topic results in confusion and delay. The KIPPAR model provides a tangible starting point for a knowledge management initiative. Limiting the scope of a knowledge retention policy will make the effort easier to accomplish and the consistency of thought and approach will allow distinctive efforts to be integrated at a later time. It provides you a way to bypass the time, effort, and cost required to attempt this task by trial and error.

chapter ten

Implementation strategies

> Effective leadership is putting first things first.
> Effective management is discipline, carrying it out.
>
> **Stephen Covey**

Welcome to my attic. This is where I collected a few ideas that may prove useful as you prepare for your KM implementation. As a consultant, I am often called on to create custom solutions for very diverse situations. I've spent my career collecting and creating methods, templates, and special tools that may be modified for most needs. That is the way I encourage you to look at this chapter. It is not a sequential checklist that ends with absolute success. Instead, read through each of these ideas and determine if it will fit your organization, where to apply it, and, most importantly, when. Although there is no precise order for these topics, I have organized them by first listing tactics that fit best when you are initiating your KM efforts and then you will find a set of operational activities.

KM initiation activities

Establish common definitions

Before you launch a KM effort in your organization, it is vital that you first make sure everyone shares a common understanding of KM's basic concepts. Since 2000, there have been such a variety of attempts at defining this emerging discipline, you will find many who think they already "get it" but are coming at this challenge with a very different perspective. Just as I did in the early chapters of this book, you will need to lay a valid foundation for your KM discussion. Specifically, you will need to do the following:

- Document why your organization is pursuing a KM implementation.
- Help people understand the difference between and significance of both explicit and tacit knowledge. Provide examples of each. If people only focus on explicit knowledge, you will be left with content management, not knowledge management.

- Identify the KM beliefs and practices that will be included in your efforts. Document specific examples that help illustrate each of these KMBPs.
- Establish a common definition for organizational knowledge. Provide tangible examples of data, information, and decision-making elements from your organization.

Define your knowledge management vision

It is important that all of the people in your organization, from top executives to bottom-level staff, understand your organization's vision for knowledge management along with what is being done to support the effort. There will be a tendency to dismiss KM as conceptual and esoteric, but there are numerous activities currently underway in your organization that already support one or more of the knowledge management processes. Begin by asking your senior leadership to articulate what they hope to achieve from a KM effort and then guide them through setting a goal statement for each of the KM processes. These enterprise knowledge goals, or EKGs, can be compiled from the answers your leaders give to the following questions:

1. *What improvements do they wish to accomplish with each KM process?* For example, why are they interested in knowledge discovery? What do they expect from knowledge transfer?
2. *What type of improvement do they hope to see by implementing each KM process?* There are four major types of improvements that should be considered. They are the following:
   - Faster: Increased speed or reduced delay.
   - Cheaper: Increased revenue or a financial saving.
   - Better: Higher quality or a perception of quality.
   - Smaller: More results with less effort, reduced space, or fewer people.

   Are they expecting faster response to client questions, are they trying to save money, are they hoping to eliminate redundancy, or is it all of the above? Whatever the motivation, your leadership needs to provide—leadership—by explaining their expectations.
3. *Who will benefit from the improvement based on the KM processes?* People need to know who the direct and indirect beneficiaries are for implementing a specific KM process. Is this effort client oriented or intended for internal value?
4. *What will each beneficiary be able to do once the KM process is implemented that they are not able to do today?* How will the KM process enable this capability?

By having your leadership formalize the EKGs associated with each of the KM processes, you should be able to create a series of very articulate declarative statements that express your management's intent for knowledge management. From these you can help them craft a meaningful vision statement that will guide the total KM effort.

It may seem awkward, but I believe it is also important to personalize a version of the four KM beliefs. If your leadership cannot declare their support for each of these basic KM beliefs, I suspect you will have a difficult time convincing your people to share knowledge, commit to learning, honor best practices, and participate in communities of practice.

Once you have senior management's expectations for KM in hand, ask each of the subgroups who are within the scope of this KM vision statement to identify specific activities they are doing that support the goal statements established for each of the KM processes. As you document the group knowledge objectives, or GKOs, it may surprise you to see how many isolated activities are all ready underway that offer direct contributions to one or more KM processes. In fact, many of the people who have supported these efforts will be thrilled and relieved to be recognized for their contributions. Most significant about this list will be showing how some of these localized efforts may be linked for greater benefit. You may discover different groups pursuing redundant strategies where greater value could be derived by having these groups join forces.

Last, ask each organizational unit to identify future activities on their radar screens that they expect to complete within the next year. Again, this will help establish continuity and sharing.

The enterprise knowledge goals and group knowledge goals should be consolidated into a KM vision statement. This document will allow your management to see the practical side of KM by removing mystery and misunderstandings. It will also help set realistic expectations as you launch a formal KM project. I have provided the template I use for this document in Appendix A. I also provided additional information on creating this document in Chapter 6.

Assess organizational beliefs

A successful KM implementation may require significant behavior changes on many levels. One of the most significant concerns that accompanies KM implementations is resistance to sharing individual knowledge or reusing existing knowledge. People in your organization may fear that sharing their knowledge makes them less valuable. Others will simply opt to re-create knowledge rather than even look for existing artifacts.

Psychologists tell us that people often resist change due to strongly held beliefs, even if they are invalid. If you hope to facilitate behavior change, you must understand what the individuals in your organization believe about the behaviors you are asking them to adopt. Part of this examination also includes how peer groups perceive the behavior and then determine if it is even feasible to perform the behavior.

The easiest way to understand these factors is to perform a formal assessment of beliefs as they apply to the KM beliefs and processes (KMBPs). A legitimate assessment should be professionally prepared and administered by industrial or workplace psychologists trained in persuasion theory. Consider distinct assessments for your senior leadership, management, and practitioners. This will enable you to see significant disconnects that may exist among these groups. Lastly, a formal assessment should result in a plan of action to either change the beliefs that prevent change, correct the cause of invalid beliefs, or alter the environment causing a conflict of beliefs.

For example, you will have a very difficult time convincing people to share private knowledge if they believe it makes them expendable and their peers are disinclined to participate. Their willingness to share is even further weakened if they believe their efforts will not be respected.

Encourage communities of practice

As identified earlier, communities of practice, or CoPs, are informal groupings of people who share a common area of interest. Your KM implementation will require a host of these groups. A good way to start comes by creating a CoP for your KM implementation. Use the KM vision statement to identify organizational units that are already making progress implementing tangible activities that support the KMBPs. Invite champions from those groups to help provide KM guidance to the total organization. Participants in this CoP will not only provide valuable advice, they will also communicate a common vision back to their home groups.

Every set of methods or best practices is a candidate for a CoP as are distinct types of artifacts. The key isn't the number of CoPs in your organization but what they are empowered to do. To make these groups meaningful, they should be encouraged to do the following:

- Set performance standards related to their specific area.
- Identify required and recommended support activities.
- Define required and recommended templates along base metadata standards.
- Select training for their area of interest.
- Select software tools for their area of interest.
- Certify sample results from their area of interest.
- Publicize successful applications of recommended activities.

Launch your KM effort as a project

Projects are the formal mechanism for change in any organization. Implementing knowledge management as a formal discipline in your organization may be one of the most challenging and valuable projects you work on in your career. It will include significant discovery activities, a changing scope, and a laundry list of stakeholders. A KM initiative offers substantial organizational benefit, will demand financial investment, and will encounter risk. At some point, you will see your KM efforts become more operational with smaller incremental projects to make improvements or corrections. Your initial KM efforts, however, should be treated as a formal single-time project, as follows:

1. *Create a written project charter:* A project charter is an agreement between the requesters and creators of project deliverables. It identifies the expectations for the project along with known conditions that must be accommodated. A charter defines scope and identifies the roles and responsibilities for the people who will help make the project successful. Once created, your project charter should be presented to senior management for their approval. When change is required or requested, the project charter is revised to document this shift in direction.
2. *Forecast a base project plan:* A project charter sets the boundaries for project plans by establishing how the project started and where it is going. Plans fill in the gap by describing the tactical strategy to complete the project. Project plans should be documented first as a high-level base plan that provides the general direction for the project along with meaningful review locations where project leadership can meet with senior management to make sure the project is progressing as expected. More detailed plans for the project team may then be created on an incremental basis by listing specific assignments.
3. *Follow a repeatable project life cycle:* Your organization should already embrace a formal process for project management that addresses initiation, execution, and completion for each project. Project *initiation* includes all of the activities needed to launch your efforts, including the creation and approval of your initial project charter and project plan. Once that has been approved, project *execution* is the cyclical process of producing a detailed plan for the work your team will perform over the next 30–45 days, controlling the work to make sure it is being accomplished and then assessing the quality and success of the work. When the project completes the goals and objectives stated in the project charter, project *completion* activities provide a fertile ground for lessons learned.

One benefit from launching your KM efforts as a formal project will be to use it as an example of how project-related knowledge artifacts may be created, retained, and used. You gain credibility anytime you practice what you preach.

Build a knowledge portal

One message from organizations with successful KM implementations is clear: knowledge management must become a natural component of the work environment. It helps to conduct business normally while finding ways to distill organizational knowledge along the way. One of the best ways to do this is to create a *knowledge portal*. A knowledge portal is the entry point for storing new knowledge or finding existing intellectual assets. It should be the mechanism all members of your organization use to access electronic resources. Once logged in, your employees should find well-organized knowledge resources at their fingertips in a way that is easy and logical to navigate.

The key to this knowledge portal is to group knowledge assets in a way that makes sense to the people in your organization. That is why I stressed the creation of a knowledge inventory or knowledge retention policy (KRP). Each enterprise area addressed by a KRP should be a distinct section on the portal that will take you to a "home page" for the enterprise area. On that page can be the knowledge areas leading to knowledge topics that support this enterprise area. You may also want to create distinct windows for recognized best practices and templates that support this specific enterprise area.

To make the knowledge portal even more useful, consider adding features to the home page that take your staff to information assets that are more general in nature. For example, consider links to the following:

- A people search for knowledge experts within your organization. This should be searchable on name or skill area. Consider using a special "ask the expert" format that will be used to locate people inside and outside your organization who have pertinent information they are willing to share.
- A project portal that lists all of the active projects being performed within your organization. This list should provide links to important project management artifacts such as the project charter, project plans, issue logs, change requests, and status reports. It should also identify the key people participating in each project. More discussion about project portals can be found in Chapter 11 when I focus on KM solutions.
- Operational processes and resulting documents.

- Organizational processes and forms needed to support general employee activities.
- Industry-related news and updates.
- News feeds that display information that might affect your organization or customers.

This strategy will allow you to build up your knowledge portal, one enterprise area at a time. Create the basic structure to support your initial KRP and add to it. Make the repository technology you use for this portal transparent to your users. You don't want them to have to acquire special technology skills to navigate the portal. Keep it simple.

You may also want to consider linking employee login, common software products, and even e-mail access to the knowledge portal. When fully implemented, this product will be the first thing people see when they start work each day and the last thing they turn off.

Create a sample knowledge retention policy

By this point, you either see the value of a formal knowledge inventory or you do not. If you do, stop thinking about it and create one. Select an area of your business that will either be easy and present little challenge, or bite off an enterprise area with known risk of knowledge loss. Either way, just do one. If your management isn't buying into the concept, select an enterprise area where you have great familiarity and will need little assistance. For example, start with your own department.

Once you have something tangible in hand, people will understand what you are talking about and may be willing to see this concept expand. If not, what have you lost? What you will have gained is a knowledge succession plan for your part of the organization.

KM operational activities

Define personal knowledge goals

Of all the topics in this chapter, this may have the most impact on creating a successful KM environment. It is practical and I've seen it work. Earlier in this chapter I suggested creating a knowledge management vision statement to help set big directions for KM. In that discussion, I recommended that you establish enterprise knowledge goals and group knowledge objectives that articulate what the total organization expects to gain from knowledge management. As the saying goes, however, "The devil is in the details." If you want KM to work for your organization, it must work at the individual employee level.

This can be done using a well-constructed plan that provides incentives and consequences around personal knowledge objectives, or PKOs. For it to work, this process must be integrated into your employee evaluation process. If you can't make that happen, don't waste your time on this. The good people who share their knowledge for altruistic purposes don't need any further motivation. However, they will welcome this program as it will recognize them for good deeds they are already doing.

This type of program is intended to let the total organization know that knowledge discovery, capture, organization, transfer, use, and retention are serious business as is their participation in communities of practice, using and contributing to established best practices and pursuing individual learning. With the help of your human relations organization, do the following:

1. Help each staff member, from executives on down, to identify one or more specific, provable contribution they can make to organizational knowledge. The intent is to find tangible ways to share something they know that is valuable to others in the organization. You may use these personal knowledge objectives to establish ways to either transfer or use organizational knowledge.
2. Connect each PKO to documented organizational directions listed on the KM vision statement. Aligning individual objectives with organizational directions will help people see how their efforts will benefit the enterprise.
3. Identify a knowledge transfer vehicle that fits best with the employee's knowledge contribution. This could include documenting a procedure, publishing a paper, conducting a training session, producing a presentation, making a video, creating an interactive product, or mentoring a new staff member. Give the employee some room for creativity here but come to a provable measurable result.
4. Provide the support resources to help the employee meet their PKOs. This may mean offering the assistance of technical writers, professional training support, or video professionals.
5. Allow employees time to meet their PKOs. Ideally, these goals should be accomplished during the normal conduct of their responsibilities. But they may require additional time to finalize the results. Have the employee estimate the time required, come to an agreement, and hold them to it. It is not reasonable to hold people accountable for achieving their PKO and not provide them with needed time.
6. During incremental reviews, follow up on the progress individuals are making on their knowledge objectives. Provide additional assistance when needed but avoid removing the objectives unless they are simply no longer meaningful.

7. Conduct a formal review of the PKOs during the employee's next annual evaluation period.
8. Provide incentives and consequences based on the successful completion of an employee's PKOs. This could include participation in a bonus pool, salary increase, or promotions. Meeting the individual knowledge goals must matter. If not, the program will soon disintegrate. Consequences may include additional training, mentoring, or even demotion.

Expect some pushback on this program as some people may wish to avoid this type of accountability. This is how you will communicate your organization's commitment to knowledge management.

Harvest knowledge assets from projects

The primary focus of the KIPPAR model is to recognize projects as a source for new or refined knowledge within an organization. Even if you are unable to implement anything else suggested by this book, here are three "project-centric" strategies you can put in place for your projects.

1. *Document affected knowledge assets:* As I've established in this book, projects touch many types of knowledge assets during the natural course of creating or modifying new products or services. When launching each project, ask team members to compile a list of affected knowledge assets they anticipate will be created, used, or revised during the project. For example, I've incorporated this element into my project charter template. In this section, I list the name of the knowledge asset, why it is significant to the organization, how the project will touch the artifact, and where it is located. I propose using "CRUD" analysis when considering how the project will affect the knowledge artifacts. For example, will your project *create* some new collection of knowledge? Maybe the artifact will simply be located and *referenced*. You may also find the project *updates* a knowledge artifact that was referenced. Finally, your project may *delete* or archive knowledge that is no longer relevant to the enterprise. Once you start this list, I think you will be surprised how quickly it grows. You may need to make some determination of significance. I use the same "vital/important/convenient" ranking here that I use in the KRP. This allows me to focus my efforts on the more significant knowledge artifacts.
2. *Review list with project owner:* One of the challenges you will face is finding sufficient time during the project to properly document and capture important knowledge assets touched by your project. Review the affected knowledge assets list with your senior management to

gain their agreement. If they concur with your assessment, they will be more inclined to allocate the time and money required to complete the job. Far too many vital knowledge assets are being lost today in the rush to just "get it done."
3. *Emphasize lessons learned:* Projects are a natural testing ground for new strategies and methods. This is where you get to try out new knowledge you or team members have acquired. At the close of your project, don't overlook the most important contribution your project can make to the organization. If you truly believe in quality, lessons learned activities are as important as any product or service you have created. This is your chance to reflect on new approaches and to create or update organizational knowledge about best practices. Ignoring this process voids the notion of knowledge management and it often dooms an organization to repeating the same mistakes. You formalize this process by adding this component to your project plan, creating a procedure to guide this postproject evaluation, building a template to capture resulting information, and then making sure this new knowledge is updated to the knowledge inventory and communicated to other project specialists.

Although implementing project-centric KM strategies will not solve the knowledge loss facing your organization, it will lay a foundation for when your leadership begins to address this challenge in a more holistic manner.

Engage contributing disciplines

The approach to knowledge management proposed by this book will benefit from the contributions of numerous disciplines and professions. To help your KM initiative take root, you will find it useful to call on a variety of groups to bring their skills to the process. Table 10.1 provides you with a starting point to consider the contributions of various specialties and organizational units.

Emphasize "management" elements of KM

The word "management" is easily associated with a multitude of business-related disciplines. In some cases, it serves only to provide a more legitimate-sounding title. For your KM implementation to remain a true "management" discipline, it must support established processes around how you plan, lead, organize, and control knowledge.

Think of *organizing* as the intentional structuring of people and materials to achieve effective decision-making and efficient operations. Organizing KM should include the following:

Table 10.1 KM Contributors

Discipline	Contribution
Library Sciences	How to create rigorous taxonomies of content and metadata. Provides foundation for understanding repository theory. Primarily focused on general/encyclopedic knowledge.
Psychology	Can explain how to encourage organizational and personal behavior changes.
Project Management	Provides a repeatable discipline for managing projects. Can be a leading source for locating new knowledge, using existing knowledge, applying best practices, and generating lessons learned.
Problem-Solving Disciplines	Exists in most professions. Many discovery and design solutions are based on detailed methodologies supported by these disciplines. Ideal source for best practices to discover and capture knowledge.
Information Technology	Provides the infrastructure and expertise required to attain and support repository products. Can design and build knowledge portal.
Human Resources	Maps the people succession plan to a knowledge succession plan. Hires people based on needed knowledge. Can help design process to include personal knowledge objectives in the employee evaluation processes.
Legal/Accounting	Validates compliance with industry and government standards. Contributes range of best practices and templates.
Operations/Business Units	Uses the knowledge created by other groups. Contributes new operational knowledge based on best practices.

- Creating a general taxonomy for organizational knowledge in the knowledge inventory that provides structure for automated and manual repositories.
- Establishing human structures to monitor and maintain the knowledge retention policy and then coordinating the projects, processes, and artifacts pillars.
- Defining a project structure prior to launching a formal initiative to implement this model.

Planning is anticipating the proper tactical steps required for a successful organizational endeavor. This should include the following:

- Articulating clear goals and objectives for establishing organizational knowledge as a valued and useful asset in the KRP's general management statement.
- Clearly defining what portions of an organization are in and out of scope for a specific knowledge retention policy.
- Anticipating the steps required to implement all of the elements of the KIPPAR model including expectations for the projects, processes, and artifacts pillars.
- Identifying an ideal sequence for the tasks that must be accomplished to implement the model.
- Assigning sufficient human and material resources to this plan to produce the expected results.
- Determining schedule expectations for the total knowledge management initiative.

Controlling provides the mechanisms needed to ensure compliance with organizational expectations. For retaining knowledge within an enterprise, this must include the following:

- Establishing quality standards, including required and suggested metadata, for all knowledge by-products of the KIPPAR model.
- Verifying that the quality standards are met.

Not as tangible as organizing, planning, and controlling activities, *leading* is the act of motivating and influencing people. Strong organizational leadership will be required of senior management to implement KM, as follows:

- Continually communicating the asset value of organizational knowledge.
- Establishing an environment and corporate culture that encourages knowledge sharing and reuse.
- Participating in the identification of the knowledge areas and knowledge topics for a KRP.
- Providing the tools and resources to create and maintain the knowledge retention policy.
- Providing the tools and resources to support effective repositories.
- Demanding the implementation of valid project management processes.
- Encouraging the identification and use of highly repeatable processes that produce consistent artifacts.

- Requiring the use of established organizational knowledge in future endeavors.

Conclusion

It is hard to know when to call this chapter "done." As more organizations implement serious knowledge management environments, we will continue to learn new things about these topics. So be creative. I will.

chapter eleven

Knowledge management solutions

> The most exciting breakthroughs of the 21st century will not occur because of technology but because of an expanding concept of what it means to be human.
>
> **John Naisbitt**

So many of the writings on knowledge management seem to begin and end with technology implications and how to create a massive centralized, integrated solution. By this point in the book, you are already familiar with my concerns over "enterprise" solutions and instead, recommend smaller enterprise areas and a project-centric strategy. I often find that organizations have already purchased their technology components prior to understanding the business problem or diverse needs for knowledge. This is yet another attempt to buy a solution to the knowledge gap, one that has little chance of success. Yet, technology clearly has its place in the KM discussion so I've chosen to close my book with thoughts on how to build meaningful implementations for your knowledge assets.

Without question, what motivates much of the interest in knowledge management is the opportunity to create technology-based solutions to a series of pressing problems. In fact, the very option of creating elegant KM solutions makes the effort of uncovering organizational knowledge worthwhile. But what type of solution are you looking to create? If not careful, the drive to create "something" will overwhelm good design and leave you with the same disappointment shared by so many others. This is not a simple "buy it" decision, regardless of what the repository and software product vendor wants you to believe.

As I meet people who have created comprehensive KM solutions, I hear a consistent theme. They tell me that they could have accomplished their ultimate objective with any of the major repository products on the market. Furthermore, the actual software platform used to build KM portals should be transparent to the ultimate user. If your knowledge workers are required to understand the technical interworkings of a repository product in order to use your KM solution, you will encounter significant resistance and limited use. It will also require constant retraining every time you change versions of the solution technology.

Instead, come at this as if you were creating a "black box." Black boxes are solutions where the ultimate user has no idea, or concern, over how a particular solution was created or how it works internally. My car, computer, and the clothes I wear are all black boxes to me. I recognize that great care and work are required to assemble them, but I have little knowledge of how they were designed, assembled, and distributed. I just know that they work, most of the time, and when they break, someone knows how to fix them.

If you want widespread use of your KM solutions, they should be created with a similar mindset. Your users should know little, and care even less, what repository software product your KM solution is based on or the way the product was designed. Users should simply find the use of organizational knowledge a natural part of their daily routine. To achieve such lofty design goals, you will need to focus on two features of any KM solution you create: usability and functionality.

Functionality

Functionality has to do with providing the right information in a cohesive coherent flow that seems natural to the business user. Common concepts should flow together to make up an integrated whole. Users should not have to hop from place to place trying to find the information they need. If you simplify the information seeking, you will have created a solution that people actually use.

The key to functionality is defining meaningful groups of organizational knowledge based on the knowledge retention policy defined earlier in this book. Structuring your knowledge around well-thought-out enterprise areas, knowledge areas, and knowledge topics prepares you for this feature. Your job remains to package the results by either consolidating multiple enterprise areas into a single portal, or even partitioning it out to the knowledge area level.

The easiest way to determine this outcome is to engage with the various communities of practice that represent the most likely users of specific information. They can give you practical insight into how they plan to apply and refine the retained knowledge over time. Keep in mind that the people who helped you define this organizational knowledge may not be the same people who will actually use it. Let subject matter experts tell you what needs to be in your knowledge solutions. But let the end users define for you how they will access the material. Our technology world is full of unusable, poorly designed solutions that failed to consider this distinction.

Usability

In addition to functionality, you must also consider usability factors that reduce unneeded barriers to the adoption of an ultimate solution. Usability encompasses many factors, from making sure the knowledge is available to the end user when and where it is needed, to the actual arrangement of the material. A premium must be placed on making sure our knowledge users can take the knowledge where they are most likely going to need it. Increased barriers to accessing organizational knowledge reduce the potential that the right information will be used in appropriate situations. If this means having internal organizational knowledge assets available at client locations, make sure your knowledge is highly portable. Waiting until your outside people can get back to the security of your facilities could also open the door of opportunity to a competitor and damage your business.

Part of your usability decisions will be based on whether the ultimate knowledge consumer is internal or external to your organization. As proven by the explosion of the Internet, a key to growth of your business is getting knowledge about your organization's products and services into the hands of current and prospective customers. But you cannot rest on the laurels of having a well-designed website. Having a cleanly designed catalogue was vital to mail-order businesses in the mid-1900s. Much later, publishing material to slick CDs and video media was another progressive step. But none of those solutions carries significant customer interfaces today. As I discuss in this chapter, I anticipate another major shift in technology that may make even the best designed Internet or intranet site appear clunky and unusable.

There are many theories, books, and workshops that attempt to guide your technologist toward more "usable" solutions. Most of these approaches, however valid, are constructed on the concept of "mass." Technology application created for the masses must consider what most people consider usable. This applies to the common software you use for word processing all the way to specialized solutions you use to manage customer accounts. They were not created for just you, but all of you. As a result, what might work for others doesn't always work for you.

In the introductory chapter to this book, I discussed what I believe are the four distinct generations of knowledge management thought. By this point, I hope you can see more clearly how the *managing organizational knowledge* approach (third-generation KM) helps you get to a well-structured view of your deep knowledge. But now we need to consider the fourth era, or the *personalized* view of a KM solution. Then we can move on to consider the benefits and risks of *departmental* and *enterprise-wide* portals.

Personalized knowledge apps

Society has clearly moved beyond the massification influence of the industrial age. We are no longer satisfied with the same answer as everyone else, from the styles we wear or the food we consume (how many choices do you want for your hamburger), to the information we need. Massification is being replaced by either customization or, more likely, mass customization. Witness the explosion of apps that support the latest generation of smartphones and tablet computers.

These miniature applications target specific needs for specific subgroups of individuals. Each person is then empowered to select exactly the tools needed for a specific situation and configure them on a personal level. A hundred people could have the same apps, but when arranged in different ways, they represent a customized solution.

As with the apps, I believe this trend is the future for information users. The objective of knowledge architects and designers must be to create clusters of meaningful, deep organizational knowledge. Working with technologists, we need to construct miniature KM portals that our ultimate users may select and load to their personal devices. As our users' needs change, so will their selection of the knowledge apps.

Distinct communities of practice should examine the multitude of app choices and recommend a core set of knowledge clusters that will help their group perform standard tasks. Keep in mind that some of these will be from internal sources where others may come from external marketplaces. It would then be an individual choice to add or subtract from that group. Organizational consistency will still be maintained by guiding the content, internal taxonomy, design, and construction of these knowledge applications.

You might anticipate resistance to this transition to personal knowledge applications from some information technology organizations, which have long specialized in creating large enterprisewide solutions with standard software products. Such solutions have helped fuel organizational consistency and global integration. Information technology organizations rely on using specific hardware and software components and this move to personalized application will raise legitimate security and support concerns. However, having been part of that industry since the early 1970s, I also know that such resistance can stem from an unwillingness to stay current with a rapidly changing technology landscape. I have no desire to paint an entire industry with the same brush, but you may need to find external subcontractors to help with the construction of your knowledge apps. Allow your IT organization to define the ground rules, including the repository products and general technology architecture. Then acquire the resources of independent app developers. And if you think these people will be hard to find, you haven't watched the ever-exploding catalogue of app choices on the market today.

Do not confuse use of these knowledge apps with their coordinated creation. Once a significant knowledge app has been created, it is vital to assign it to the appropriate community of practice where it should be regularly reviewed and updated. Lessons learned from individual or group use will be an invaluable guide as you grow the need and demand for this collection of knowledge.

Organizational portals

So your organization is not quite ready for the "app generation" with your knowledge assets? You still have the option to create knowledge portals that support specific units of your organization or even the total enterprise. As these structures have been in place for at least two decades, my suggestions are intended only to make them more usable.

When creating knowledge portals for a specific subset of your organization try to avoid allowing your current political structure to influence your knowledge taxonomy overly. As I discussed in the knowledge retention policy, our initial structures often follow familiar patterns based on how an organization is structured. The clustering of knowledge, however, often exceeds the use by just one part of the organization. It is common to find that needed information has been locked away by one department in a "private" location, resulting in content redundancy and inconsistency.

Even if you are committed to creating a single point of entry to your organization's knowledge, make the experience individual. Provide your staff and customers with the option to configure the knowledge assets in a manner that fits their use. How they traverse the structure should be insignificant. Arriving at the correct destination is what you want.

Highly useful enterprisewide knowledge portals do the following:

- Provide quick access to commonly used forms and templates.
- Allow access to external information sources that keep your staff current on products, clients, competitors, and markets. These often come in the form of feeds from news or industry sources.
- Integrate each person's work activities with the other administrative elements that concern the employee. For example, make the portal the entry point for calendars, e-mail, and video-conferencing services. Provide an area in the knowledge portal where people can access personal information related to their employment such as health plans, investments, and other benefits.
- People want to know about the new things happening in their world, especially new products, services, or procedures that may

affect them. A link to a project portal (described below) will help keep people informed about new directions for the organization.
- Integrate the use of appropriate social media into your portal. Create comprehensive profiles of your internal, and even some external, knowledge resources so people can search for an expert. Provide discussion threads and the opportunity to subscribe to communication updates. These might include bulletins from the communities of practice each person finds of interest (more on social media below).

Creating a highly effective organizational portal will be a never-ending proposition. Technology changes, along with shifts in needed knowledge assets, will make this a continuing challenge.

Project portals

All thriving organizations rely on new products and services for both their internal and external clients. We deliver on this demand through special projects or initiatives. The incessant demand for new projects taxes the resources of most organizations. But instead of making the process easier through repeatable processes, common templates, and clearly defined communication processes, too many organizations treat all this as some type of art form where each person re-creates the wheel under the name of "innovation." True innovation should be experienced in the ultimate products of a project, not in the project management process.

Helping organizations formalize a process for project management has consumed me for decades. One by-product, however, has direct implications on a portal strategy, a project portal. The structure of your organizational knowledge repositories will reflect the actual intellectual assets your organization wishes to keep, however, project portals are easier to define. Applying consistent formal methods to this discipline provides valuable insight into the most commonly created and requested by-products from a project. When creating a repository for project content, be it a three-ring binder, shared drive, or a technology-based repository product, I recommend the following:

1. Creating a common entry point for all projects in your organization.
2. Establishing a distinct presence for each individual project.
3. Distinguishing between administrative and functional content for each project.
4. Transitioning meaningful project artifacts to an operational state.

Chapter eleven: Knowledge management solutions

Common entry point

Where feasible, there should be a single point of information for all projects within your organization. It should list active projects along with those that are recently completed and projects awaiting approval to move forward. A project archive would allow current project managers to mine previous content for samples and content.

For the most part, the project portal should be an open book within an organization. There may, however, be a few projects that must remain highly confidential. If you have some of those efforts underway, you may need a way to hide them from the more public view.

When creating the list of projects in the project portal, provide sufficient information to clearly identify the name of the project and the name of the project manager. As the most commonly asked questions about a project center on a current status, some find it helpful to either add a status field or even a one-sentence status statement with a link to "more information." The listing for a specific project should provide a direct link to a site with a collection of information specific to that project.

If you have a large number of projects underway in your organization, consider giving each user the opportunity to create a tailored view of the project list. For example, you may want to limit the projects displayed by the project manager, project owner, participating organization, or some combination of these elements. Should you find a key project player who is still uncomfortable using browser technology, consider adding a link on his or her browser to a relevant customized view of project information.

In addition to a list of projects, a project portal should also include links to recommended templates for common documents along with suggested or required processes such as estimating methods, scope change procedures, or financial worksheet preparation. It is also wise to capture good examples of project deliverables found during your postproject assessments. If your organization has created one, you may also want to include a link to the general project life cycle that explains expected project management activities.

Distinct presence

Every project on your master list should have a distinctive presence in your project portal. Consider this a "home page" that is dedicated to each individual project. This location should be the launching pad for learning more about the project management deliverables being used to guide the project along as well as the more technical by-products being created. Links to important documents and discussion threads should also have a home here.

Consider creating a template site that may be used to set up each new project and then give the project manager and team the option of adding or subtracting content as needed by their type of project.

Establishing a distinct presence for each project allows your total organization to quickly locate and review what is "new." It gives all project players a common place to find information. It also gives other interested employees an easy way to view the projects that will shape your organization's future. Furthermore, these project sites remove any concerns by your people that they are being kept in the dark. It puts them in the driver's seat with only a few clicks. I like to add subscription options where people who have a dedicated interest in a project can be notified of new or revised content by e-mail.

Administrative versus functional content

Projects produce massive amounts of content. Although creating a distinct presence for your project in the project portal gives you the logical place to access this information, the variety and volume of content makes it wise to think through a logical file structure where project team members may load and search for significant materials. Your initial act should be to distinguish between administrative and functional content created during the project.

Administrative project documents are those used to manage the effort. They tell you about the effort. In this area, I typically start with folders for my project charters and agreements, project planning components, issues log, change requests, and status reports. Why? Because I know that valid project management processes will produce multiple versions for most, if not all of these documents. Instead of using my structure, a better idea might be to empower your community of practice for project management to establish a common file structure that represents the content recommended for your organization. This structure should be the starting point for all projects with additional folders added for specialized content. Consistency across projects will make it much easier to anticipate where to find critical content and good examples.

Functional project documents represent the actual work products created or revised during a project. This content is determined by the type of project underway. Functional documents may include engineering specifications, drawings, designs, schematics, business processes, procedures, reports, test criteria, permits, test data, and the list goes on and on. Because of the diverse nature of this list, it is important to make the identification of *affected knowledge assets* part of your project kick-off.

Determining how to organize the functional project documents should be guided by the types of documents being created. If you find consistent project types in your organization, just as I recommended for the administrative content structure, work with the communities of practice who represent the methods and best practices used to create the functional content. They will be able to recommend a standard structure for this content. If your content is unique or you cannot identify a common structure, consider creating a knowledge retention policy for your project. The structure process in a KRP will help you design a stable file structure.

Once you have an organization of administrative and functional content completed, you may want to return to the home page for the project and add direct links to commonly requested information.

Transition artifacts

When a project comes to an end, the work with project-related content is not over. Each project manager should review all the content captured during the effort. For the most part, the administrative documentation may be archived. It was extremely useful to manage the project, but has limited long-term value. Don't throw these materials away or delete them. A lessons learned process should have already identified future improvements to your project management processes and best practices. If some of the administrative documents represent the type of content and quality desired by the organization, move them to a special area in the project portal reserved for samples.

Functional project documents were used to create the ultimate products and services from your project. A review of this content should reveal the materials that will provide long-term value to the support and maintenance of the resulting project deliverables. For example, if your project created a new technology-based product that facilitates a relationship with your customers, you may find it helpful to keep engineering specifications and design documents that led to the eventual product. Instead of just tweaking the product for refinements, these functional documents would allow future staff members to understand the underlying thinking that resulted in the product. They will be able to make more informed changes rapidly.

Any functional project content deemed to have long-term value should be transitioned out of the project repository (where it was created) to an operational repository (where it will be used). In too many case, this valuable information is left in the file structures, repository, or shared drive, created for a specific project. When needed later, the hunt begins by trying to find someone who worked on that effort who "might" remember where a specific document was stored.

As this project-derived content represents new knowledge assets for your organization, it should fit into a knowledge area and knowledge topic identified in a previous KRP for a knowledge domain of your enterprise. If this doesn't exist, this may be the time to build this permanent inventory as a way to ensure the knowledge asset is well managed over time. All remaining functional content should either be deleted or archived.

Conclusion

> If you have knowledge, let others light their candles with it.
>
> **Sir Winston Churchill**

There is simply no tidy way to wrap up a book like this. It feels more like a "pause" than an ending. For I am certain these ramblings of mine do not represent the final word on knowledge management. Although I have attempted to deliver to you the most current understanding I have on these topics, every time I use the processes I've described, I find something new. I continually update my processes and templates, and I doubt that process is anywhere near complete, for me or for others.

My hope is that I've stimulated new thought in your mind and generated discussions within your organization. By writing this book, I believe I have lived the very intent of knowledge management. This includes discovering new ideas through research and application and then capturing those observations in written word and in templates. I attempted to organize this content so you could see a coherent flow and find needed content later when you wished to use it. I trust these words managed to transfer this knowledge to you so that these ideas may be retained over time.

I have long been reluctant to write a book such as this. A former mentor set a very high standard by decrying the number of books that are simply filled with the "same old stuff." He observed that "new" books often restated ideas previously published, and as a result, added little new value. I waited to write my first book until I had something unique to say. I believe I have brought new observations on established thought and introduced a number of new concepts along the way.

The only true indication of the success of this book is if it produces meaningful valuable change. The change I hope to see will be people and organizations declaring a strong vision for knowledge sharing and reuse. Organizations will retain vital capabilities long after the original experts are gone. New employees will learn to stand "on the shoulders of giants" without feeling the need to constantly reinvent the wheel. Organizations will have global access to established proven knowledge using appropriate technology. And each time a major project comes to completion, it will be mined for the new knowledge it has unearthed.

The ultimate impact of this book will be on you. Has this content changed the way you think or act? Has it inspired you to create a culture of knowledge sharing in your personal sphere of influence? By helping others achieve their full potential, you truly are helping yourself. Knowledge—pass it on!

Although we may have never met, you have my sincere thanks for reading this book. I am easy to find through my website (www.TryonAssoc.com) and I welcome your comments.

Tetelestai!

Appendix A: KM Vision Statement*

---

* PDF versions of the templates shown in the appendices may be found at www.TryonAssoc.com/news/index.asp#templates.

KNOWLEDGE MANAGEMENT VISION STATEMENT
<Org>
<Enterprise Area>

KM VISION: <Add text here...>

KM PROCESSES	Knowledge Discovery	Knowledge Capture	Knowledge Organization	Knowledge Use	Knowledge Transfer	Knowledge Retention
Executive Knowledge Goals (EKGs)	<Org> endorses the use of formal methods Knowledge Discovery by...	<Org> facilitates the capture of knowledge by...	<Org> will organize captured knowledge using formally defined structures and technologies that make content more easily searchable.	<Org> encourages the use of captured organizational knowledge by...	<Org> actively encourages the transfer of organizational knowledge using proven mechanisms that include...	Knowledge will be retained in <Org> as it is discovered, captured, organized, transferred and used. This capability will provide business continuity and accelerated financial growth.
Group Knowledge Objectives (GKOs)						
Business Unit 1 Activities/**Plans**	Identify current activities in this group that supports Knowledge Discovery. Identify planned activities in this group that supports Knowledge Discovery.					
Business Unit 2 Activities/**Plans**						

KM PRACTICES:
To support this Knowledge Management implementation, we hold the following to be true...
- Sharing personal knowledge or using established knowledge increases the value of an employee to <Org>.
- Knowledge increases through a commitment to individual and organizational learning.
- Proven best practices should be promoted and used at <Org>.
- Communities of practice are a valuable means to improve the quality of organizational knowledge.

Appendix B: KRP—General Management Statement**

---

* PDF versions of the templates shown in the appendices may be found at www.TryonAssoc.com/news/index.asp#templates.

KNOWLEDGE RETENTION POLICY
Level One Study

Version 1.0
<Date>

Created for:
<Organization>

Appendix B: KRP—General Management Statement

KNOWLEDGE RETENTION POLICY
General Statement

A *Knowledge Retention Policy* is formal written document that declares intellectual properties considered to be vital organizational assets. Similar to a list of physical assets, this document identifies an organization's intellectual assets. A *Knowledge Retention Policy* defines...

- Knowledge Areas
- Knowledge Topics
- Knowledge Transfer Mechanisms

KNOWLEDGE AREAS are groupings of organizational knowledge that are recognizable to the total enterprise. Knowledge Areas may be formal organizational units or important functions performed by a subset of the organization. The classification of Knowledge Areas is arbitrary and serves only to give structure and organization to Knowledge Topics.

RESPONSIBLE PARTY represents the person, persons or organizational unit who has the authority over and the organizational accountability for this specific Knowledge Area. They will be responsible for validating the Knowledge Topics and other attributes.

KNOWLEDGE TOPICS are recognizable collections of repeatable processes and/or data that are significant to the organization. Knowledge Topics may be organizational programs, business processes, business data and application systems. Each Knowledge Topic should be ranked as to how significant it is to the enterprise along with the status of knowledge transfer.

ORG. SIG.	MEANING	TRANS. STATUS	MEANING
V	This knowledge is VITAL to the organization. Failure to capture and transfer this knowledge will cause operational failure.	W	This knowledge is well defined and accurate. It may be transferred using established mechanisms. No further action is needed.
I	This knowledge is IMPORTANT to the organization. Failure to capture and transfer this knowledge will compromise operations.	L	Limited definition is available for this knowledge. Review and refinement is needed. Formal transfer mechanisms are needed.
C	This knowledge is CONVENIENT to the organization. Failure to capture and transfer this knowledge will reduce operational efficiency.	U	This knowledge is undocumented and no formal transfer process currently exists.

OWNERS/SOURCE indicates the person, persons, or organizational unit, process, software product or collection of data that is the authority on or the basis for this Knowledge Topic. A Knowledge Topic may have more than one Knowledge Owner/Source. This resource will be vital in capturing and transferring this collection of organizational intelligence. Provide names when possible.

KNOWLEDGE TRANSFER MECHANISMS represent the means used to capture and transfer organizational knowledge from one group of practitioners to another. More than one type of Transfer Mechanism may be used for a specific Knowledge Topic. Distinct Knowledge Transfer Mechanisms include...

- Documentation – One means to capture and transfer organizational knowledge is using physical or electronic documents. This may include all data types including text, graphics and video. This knowledge may be stored on any media including paper, video or electronic record.
- Apprenticeship – Complex, significant knowledge is often transferred using a relationship between an expert and apprentice practitioner. Apprenticeship implies a dedicated, sustained transfer process.
- Training – Organizational knowledge is often transferred using an educational process. Training may be comprised of formal education and/or specific task training.
- Mentoring – As a supplement to other forms of knowledge transfer, mentoring provides on-going benefit. Mentoring includes identifying people who are available to provide advice and assistance to someone performing a new task.
- Cross Training – Many organizations enable knowledge transfer by placing less experienced people with task experts in "job shadowing" process.
- Communications – A great deal of organizational knowledge is transferred using formal and informal communications. Formal communications include professional societies, committees, conferences, job-related websites and reference books. Unstructured communications include social networks, social events and chat rooms.

ENTERPRISE AREA identifies what will and will not be addressed in this Knowledge Retention Policy. Common scope descriptions may be the total organization, distinct operational units, internal or eternal service providers and specific projects.

Appendix B: KRP—General Management Statement

A Knowledge Retention Policy may be created in two levels. A **Level One** study identifies the Knowledge Areas and Knowledge Topics along with Topic Descriptions, Organizational Significance, Transfer Status, current Owner(s) and specific Knowledge Transfer Mechanisms that are or should be used to capture and transfer this knowledge.

A **Level Two** study expands on each of the Transfer Mechanisms by clearly defining the characteristics of this method for capturing and transferring knowledge. A Level Two study confirms the accuracy of the Level One information.

The intellectual properties represented in this document should be managed as a significant organizational asset of <Organization>. Great care should be given to create, discover, refine, capture and share this knowledge.

Authorization

_____
Name

_____
Title

_____
Date

Appendix C: KRP— Intellectual Assets Inventory[*]

---

[*] PDF versions of the templates shown in the appendices may be found at www.TryonAssoc.com/news/index.asp#templates.

KNOWLEDGE RETENTION POLICY
Intellectual Assets Inventory

ENTERPRISE AREA: This KRP Intellectual Assets Inventory identifies the organizational knowledge considered vital to the operation of <Organization>.

TABLE OF CONTENTS

<KA NAME> 6

Organizational Significance = Vital / Important / Convenient
Transfer Status = Well-Defined / Limited Definition / Undefined
Transfer Mechanisms = Documentation/Training/Apprenticeship/Mentoring/Cross-Training/Communications

Appendix C: KRP—Intellectual Assets Inventory 111

KNOWLEDGE RETENTION POLICY
Intellectual Assets Inventory

KNOWLEDGE AREA: <KA Name> <Knowledge Area description>				RESPONSIBLE PARTY: <Resp. Party>	
Knowledge Topic	**Description**	**Org. Sig.**	**Trans. Status**	**Owner/Source**	**Transfer Mechanism**

(Select and copy this table as needed for each Knowledge Area)

Organizational Significance = Vital / Important / Convenient
Transfer Status = Well-Defined / Limited Definition / Undefined
Transfer Mechanisms = Documentation/Training/Apprenticeship/Mentoring/Cross-Training/Communications

Appendix D: KRP— Knowledge Transfer Details*

---

* PDF versions of the templates shown in the appendices may be found at www.TryonAssoc.com/news/index.asp#templates.

KNOWLEDGE RETENTION POLICY
Level 2 Knowledge Transfer Mechanisms Metadata

KNOWLEDGE AREA	
KNOWLEDGE TOPIC	

(Delete or ignore unneeded Knowledge Transfer Mechanism details)

Documentation Attributes

No.	Record Name	Record Type	Location	Update Type	Revision Schedule	Access Rights	Security	Retention Term	Disposal
1.									
2.									
3.									
4.									
5.									

Record Name — Document the group name for a distinct group of records that transfer this knowledge.
Record Type — Identify the type of record for this group. Type include text, graphics, video and audio.
Record Location — Where will this group of records be physically located? This includes URLs, software products or filing location.
Update Type — Is this record a static or dynamic document?
Revision Schedule — How frequently should this set of records be reviewed and updated?
Access Rights — How may this set of records be accessed? What registration, if any, is required to gain access?
Security — What type of security is needed to protect this record?
Retention Term — How long should this set of records be retained?
Disposal — How should this record be archived or destroyed?

Appendix D: KRP—Knowledge Transfer Details 115

KNOWLEDGE RETENTION POLICY
Level 2 Knowledge Transfer Mechanisms Metadata

Training Attributes

No.	Training Name	Vendor Information	Prerequisites	Certifications	Performance Measures
1.					
2.					
3.					
4.					

Training Name — *Provide the name of a formal training program (individual course or series) that facilitates this knowledge transfer.*
Vendor Information — *Include information (URLs) that identifies the vendor, course descriptions and registration information.*
Training Prerequisites — *What previous education or experience level should be completed prior to this training?*
Resulting Certifications — *What degrees or certifications are needed to verify this knowledge transfer?*
Performance Measures — *What performance measures should be met to validate this knowledge transfer?*

KNOWLEDGE RETENTION POLICY
Level 2 Knowledge Transfer Mechanisms Metadata

Apprenticeship Attributes

No.	Attribute	Description
1.	**Apprentice Qualifications** — *What skills and/or education is needed by an apprentice?*	
2.	**Apprentice Selection** — *What are the suggested selection criteria for an apprentice?*	
3.	**Selection Process** — *What is the suggested selection process for an apprentice?*	
4.	**Apprenticeship Term** — *Recommended length of apprenticeship.*	

Appendix D: KRP—Knowledge Transfer Details 117

KNOWLEDGE RETENTION POLICY
Level 2 Knowledge Transfer Mechanisms Metadata

Cross-Training Attributes

No.	Attribute	Description
1.	Trainer Qualifications — *What are the criteria for the trainer?*	
2.	Trainee Qualifications — *What are the criteria for the trainee?*	
3.	Training Frequency — *How frequently should the cross-training sessions occur?*	
4.	Training Term — *How long should each cross-training session last?*	

KNOWLEDGE RETENTION POLICY
Level 2 Knowledge Transfer Mechanisms Metadata

Mentoring Attributes

No.	Attribute	Description
1.	Mentor Qualifications — *What are the criteria for a mentor?*	
2.	Mentor Commitment — *How much total time and incremental time will be required of a mentor? What is the total length of a mentor relationship?*	
3.	Mentor Recognition — *What types of recognition or reward are available for mentors?*	

Appendix D: KRP—Knowledge Transfer Details 119

KNOWLEDGE RETENTION POLICY
Level 2 Knowledge Transfer Mechanisms Metadata

Communications Attributes

No.	Communication Name	Communication Type	Information Source	Contact	Frequency
1.					
2.					
3.					

Communication Name — *What type of formal communication process is recommended?*
Communication Type — *Professional organization, Committee, Publication, Website, Social Network...*
Information Source — *URL or address to obtain additional information.*
Communication Contact — *Name of person to contact about this communication.*
Communication Frequency — *How often this communication occurs or how often it should be reviewed.*

Index

A

Administrative *vs.* functional content, for project portals, 96–97
Affected knowledge assets, 96
Apprenticeships, 26, 64
    completion, 64
    length, 64
    qualifications of apprentice, 64
    selection of apprentice, 64
    documentation of, 116

B

"Boomer" retirements, 6–7
Business complexity, 8

C

Common entry point, for project portals, 95
Communications, 65–66
    contact, 65
    frequency, 65
    information source, 65
    name, 65
    type, 65
Communities of practice, 14, 19-25, 38, 69, 72, 77-78, 82, 90, 92, 94, 97
Compliance, 38
Consistency, 38
Continuity, 38
Cross-training, 65
    completion, 65
    frequency, 65
    length, 65
    trainee qualifications, 65
    trainer qualifications, 65
"CRUD" analysis, 83

D

Data, 2, 33, 41–44, 46, 48, 52, 76
Decision making, 2, 33, 42, 44–46, 48, 52, 76, 84
Deliverables, 68, 72, 79, 95, 97
Distinct presence, in project portals, 95–96
Document and records management (DRM), 73
Domain of effort, 72
Domain of study, 50-51, 72
Downsizing, 7

E

ECM. *See* Enterprise content management (ECM)
Enterprise content management (ECM), 10, 26
    KM and, 63
    KM *vs.*, 73–74
Executive knowledge, 29
Explicit knowledge, 31, 34-37, 39, 41, 44, 52, 54, 67-69, 73, 75
    converting implicit knowledge to, 36–37

F

Functionality, KM solutions and, 90

G

Global knowledge, 32–33

I

Implicit knowledge, 31, 34-37
    conversion to explicit knowledge, 36–37
Individual knowledge, 15, 17, 29, 32, 77, 83
Individual knowledge goals, 83. *See also* Personal knowledge goals

121

Information, 2, 33, 1–46, 48, 52, 76
Intellectual assets inventory, 109-111. *See also* Knowledge assets inventory

K

KIPPAR model, xxi, 3, 67–74, 83, 86
   advantages, 71
   KRP and, 68
   projects pillar, 71–72
   processes pillar, 68-70, 72
   artifacts pillar, 68-69, 72, 85-86
   repository products, 69, 71-73, 85, 89, 92
Knowledge, organizational importance of, 53
Knowledge areas, xi, 49–52, 57-59, 69–70, 73, 80, 86, 90, 97, 105, 107, 111
Knowledge artifacts, 3, 10, 22, 67, 80, 83
Knowledge assets inventory, 49–54. *See also* Intellectual assets inventory
Knowledge capture, 22–23, 27, 38, 57, 102
Knowledge characteristics, 37–38
Knowledge expert(s), 53
   with organizations, 4
Knowledge gap, 46
Knowledge management beliefs and processes (KMBP), 2, 13, 20-21, 29, 31, 33, 67, 76, 78
Knowledge management (KM)
   activities, 75–87
      initiation, 75–81
         assess organizational beliefs, 77–78
         build knowledge portal, 80–81
         create sample KRP, 81
         define vision, 76–77
         encourage communities of practice, 78
         establish common definitions, 75–76
         launch effort as a project, 79–80
      operational, 81–87
         define personal knowledge goals, 81–83
         engage contributing disciplines, 84
         harvest knowledge assets from projects, 83–84
   contributors, 85
   defined, 2
   ECM and, 63
   ECM *vs.*, 73–74
   generations, 91
   implementation, 9–10
      aligning organizational efforts for, 9
      capturing metrics for, 10
      defining and supporting set of beliefs for, 20
      displaying tangible results for, 10
      engaging management in, 10–11
      establishing consensus for, 20
      recognizing and attributing success of, 11
      start of, 10
   initiation activities, 75–81
      assess organizational beliefs, 77–78
      build knowledge portal, 80–81
      create sample KRP, 81
      define vision, 76–77
      encourage communities of practice, 78
      establish common definitions, 75–76
      launch effort as a project, 79–80
   launching, 79–80
   management elements of, 84–87
   operational activities, 81–87
      define personal knowledge goals, 81–83
      engage contributing disciplines, 84
      harvest knowledge assets from projects, 83–84
   role of senior management in, 84–87
      as controller, 86
      as leader, 86
      as organizer, 84–85
      as planner, 86
   solutions, 89–98
      functionality as feature of, 90
      usability as feature of, 91
   vision statement, 9, 20, 101–102
Knowledge opportunities, 8–9
Knowledge organization, 23–24
   importance, 53–54
   rationale, 38–39
      compliance as, 38
      consistency as, 38
      continuity as, 38
      increased understanding as, 39
      lessons-learned as, 39
      quality control as, 38–39
      standardization as, 38
Knowledge portal, 10, 80–81, 91. *See also* Project portals
Knowledge retention, 26–27
Knowledge retention policy (KRP), 47
   activities, 56–59
      define enterprise area, 56

Index

define KM vision statement, 56–57
describe knowledge topics, 58
determine implementation, 58–59
propose knowledge areas and assign knowledge topics, 57–58
review completed KRP level 1, 58
review with senior management, 58
apprenticeships, 64
    completion, 64
    length of, 64
    qualifications of apprentice, 64
    selection of apprentice, 64
communications, 65–66
    contact for, 65
    frequency of, 65
    information source, 65
    name of, 65
    type of, 65
components, 48
cross-training, 65
    completion of, 65
    frequency of, 65
    length of, 65
    trainee qualifications for, 65
    trainer qualifications for, 65
documentation, 62–63
    access rights to, 62
    disposal of, 63
    file naming standards for, 62
    location of, 62
    record name for, 62
    retention term for, 62
    revision responsibility for, 62
    revision schedule for, 62
    security for, 62
    technology for, 62
    type of record for, 62
    update type for, 62
function, 47–48
general management statement of, 48–49, 103–108
intellectual assets inventory, 109–112
justification, 47
knowledge transfer details, 113–119
level 1, 47–59
level 2, 61–66
mentoring and coaching, 65
    commitment of mentor, 65
    qualifications of mentor, 65
    recognition of mentor, 65
sample, 81
training, 63–64
    certification of, 64

    costs of, 64
    description of, 64
    duration of, 64
    name of, 63
    performance measures for, 64
    prerequisites for, 64
    vendor, 64
Knowledge topics, 51–52
Knowledge transfer, 56–59
    details, 113–119
    mechanisms, 26, 54–56
        apprenticeship as, 55
        communication as, 56
        cross-training as, 55–56
        documentation as, 55
        mentoring and coaching as, 55
        training as, 55
Knowledge use, 24–26
KRP. *See* Knowledge retention policy (KRP)

L

Leadership, defined, 86
Lessons learned, 16, 37, 39, 68, 72, 79, 84-85, 93, 97

M

Managing Law Enforcement Initiatives (MLEI), 51
Managing organizational knowledge, xviii, 2, 67-74, 91
Market globalization, 8
Mentoring and coaching, 65
    commitment of mentor, 65
    qualifications of mentor, 65
    recognition of mentor, 65

O

Organizational beliefs, assessing, 77–78
Organizational knowledge, 31, 33
    employees as custodians, 4
    KIPPAR model, 67–74
    managing, 2
    measuring value, 4
    rationale for capturing, 38–39
    significance, 5
Organizational learning, 16–17
Organizational portals, 93–94

P

Personal knowledge goals, 15, 20, 81. *See also* Individual knowledge goals
Personalized knowledge apps, 92–93
Problem-solving disciplines, 85
Project(s)
    improvement, 71
    management, 85
    scope, 71
    transformational, 71
Project-centric strategies, 83–84
    documenting affected knowledge assets, 83
    emphasizing lessons learned, 84
    reviewing list with project owner, 83–84
Project portals, 94–98
    administrative *vs.* functional content for, 96–97
    common entry point for, 95
    distinct presence in, 95–96
    transition artifacts of, 97–98
Projects pillar, 71–72
    features, 72
        deliverables, 72
        domain of effort, 72
        domain of study, 72
        project scope, 72

Q

Quality control, 38–39

R

Repository products, 72–73
Return on investment, 9–11

S

Senior management, 4, 58, 83
    articulation of KM beliefs by, 20
    expectations for KM, 28
    KRP level 1 review, 58
    leadership role, 86
    project charter review, 79
    values, 14
Single-time efforts, 2
SME. *See* Subject matter experts (SME)
Standardization, 38
Subject matter experts (SME), 58

T

Tacit knowledge, xxi, 31, 33–36, 38, 41, 45, 47, 53-55, 59, 61, 63, 73, 75
Tagging, 22
Technology advances, 8
Training, 63–64
    certification, 64
    costs, 64
    description, 64
    duration, 64
    name, 63
    performance measures, 64
    prerequisites, 64
    vendor, 64
Transfer status, 54
Transformational projects, 71
Transition artifacts, of project portals, 97–98

U

Unforced resignations, 7
Usability, KM solutions and, 91

About the author

Chuck Tryon is an educator, practitioner, consultant, and author.

Since the early 1980s, Chuck has created dozens of papers and workshops on Knowledge Management, Project Management, and Business Process Engineering. Thousands of professionals from many of the largest organizations in the United States, Canada, and Western Europe have attended his workshops. Chuck is frequently invited as a featured speaker for conferences and professional society meetings.

Chuck holds an undergraduate degree in Business Administration and a master's degree in Knowledge Management. His research includes identifying ways to improve knowledge worker productivity and designing numerous advanced project-centric strategies. He is also the cofounder of the Knowledge and Project Management Symposium.

He has been employed as a Knowledge Manager and Project Manager in the petroleum and healthcare industries.

Chuck and Tresa live in the Tulsa, Oklahoma area. They have two daughters and three marvelous grandchildren. Chuck's hobbies include golf, photography, and SCUBA diving. Additional information is available at www.TryonAssoc.com.